Fire Alarm
Signaling Systems
Second Edition

Fire Alarm Signaling Systems

Second Edition

Richard W. Bukowski, P.E.

Fire Research Laboratory

National Institute of Standards and Technology

Robert J. O'Laughlin, P.E., C.S.P.

Fire Protection Engineer

A joint project of:

National Fire Protection Association
Quincy, Massachusetts

Society of Fire Protection Engineers
Boston, Massachusetts

This book has not been processed in accordance with NFPA Regulations Governing Committee Projects. Therefore, any material in it should not be considered the official position of NFPA or any of its committees.

Colophon
Product Manager: Jim Linville
Copyediting: Hilary Davis, Susan Merrifield
Technical Illustration: George Nichols
Interior Design: Joyce Weston
Cover Design: Cameron, Inc.
Index: Hagerty and Holloway
Text Processing: Marilyn Lupo
Composition: Cathy Ray, Claire McRudin, Nancy Maria
Production: Donald McGonagle

FASS-94
ISBN: 0-87765-399-2
Library of Congress No.: 94-066010
Printed in the U.S.A.

First Printing, May 1994

Contents

PREFACE

For more than 90 years the National Fire Protection Association has been working to protect people and property from fire, and part of that effort has been to serve the needs of fire alarm system designers, manufacturers, installers, and users through the publication of its codes, standards, and recommended practices, many of which apply to fire alarms.

The continued rapid evolution of fire alarm systems in the last two decades has completely overshadowed the developments of the previous 100 years. We have progressed from the early fire alarm telegraph systems to signal multiplexing; from the ringing of a bell to voice communication systems; from hardwired connections to fiber optics and wireless transmission; from unsupervised smoke detectors to addressable and analog devices that report their status; from panels that were dubbed "solid-state" because of their immense weight to today's true solid-state signaling processing panels; and from fire detectors that required enormous fires to trip them to smoke detectors that respond to only a few ounces of combustibles.

All of these changes have contributed to the fire alarm systems available today, whether they be single station residential smoke detectors or high-rise computer-controlled combination fire alarm and building management systems.

At the same time, this technical development has contributed to and been influenced by the fire alarm codes and standards used to detail the performance standards, test requirements, and installation of these systems.

Together, the National Fire Protection Association and the Society of Fire Protection Engineers have created *Fire Alarm Signaling Systems* to respond to the need for a "how-to" guide to acquaint readers with the different types of fire detection and alarm systems and the requirements for each, as set forth in NFPA 72, *National Fire Alarm Code* and the Underwriters Laboratories and Factory Mutual standards.

Fire Alarm Signaling Systems is designed to help the user intelligently apply the applicable codes and standards, and to serve as a how-to guide that acquaints the reader with the basic requirements for use of each of the fire alarm systems, and differentiates between each system. It takes up where the *National Electrical Code Handbook* leaves off by expanding upon the sections of that Handbook that discuss wiring of alarm systems.

Fire Alarm Signaling Systems is *not* intended to substitute for the signaling system codes and standards, or as a summarization of them. Rather, it approaches fire detection and alarm systems from a different perspective: its purpose is to provide information on the basic requirements of fire detection and alarm signaling systems *and* explanations of how signaling systems fit and work within those standards. A glossary of terms is provided at the end, and illustrations have been included to expand upon the material presented in the text and to assist the user visualizing actual alarm systems.

This second edition of *Fire Alarm Signaling Systems,* contains many significant changes:

- The text has been thoroughly revised and updated to reflect the significant changes made to NFPA 72 over the past six years.
- A new chapter has been added, explaining the evaluation method for fire detector placement presented in Appendix A of NFPA 72, *National Fire Alarm Code.*
- Chapter 7, Signal Notification, has been thoroughly revised, including provisions of the Americans with Disabilities Act.

Other additions to the book include data on fiber optic technology, and analog and addressable devices used in fire detection systems.

Until now, there has been no single, consolidated source for the general information provided in this book, though the need for it has been evident. The professionals who contributed to this book brought more than a hundred years of cumulative experience to the task; their contributions outline the latest advances in fire alarm signaling systems.

The future of fire detection and alarm systems is promising and will grow out of today's basic knowledge. It is hoped that *Fire Alarm Signaling Systems* will be the stimulus to take fire alarm systems to their next level of performance to meet our ever-constant need to alert people, in time, for safe evacuation and prompt fire fighting activities.

The editors would appreciate receiving any comments and suggestions that will improve the contents, as future editions of the book will be revised and expanded to keep the information current.

ACKNOWLEDGMENTS

The first edition of *Fire Alarm Signaling Systems* was named *Fire Alarm Signaling Systems Handbook*. It was prepared in 1987 as a joint effort of the National Fire Protection Association and the Society of Fire Protection Engineers. *Fire Alarm Signaling Systems Handbook* was created by a team of talented professionals lead by Charles Zimmerman, who for many years was NFPAs Fire Alarm Signaling Systems Specialist. Charlie was as much a major contributor to the first edition as he was its editor. Additionally, the following technical experts contributed an unknown amount of their time and expertise to the first edition: Jack Abbott, Joe Dronin, Peter Dubivsky, Al Heim, Vic Humm, Ted Humpel, Stan Kravontka, Bob McPherson, Wayne Moore, Crawley Parris, Marty Reiss, Jim Roberts, Bill Rogers, Joseph Scheffey, Walter Schuchard, and Max Schulman. Most of what they wrote for *Fire Alarm Signaling Systems Handbook* has been updated and reused in the second edition, and the authors and editors of *Fire Alarm Signaling Systems* wish to acknowledge their contributions.

The second edition was exclusively prepared by Dick Bukowski and Bob OLaughlin.

Special thanks are due to Hilary Davis for her fine copy editing job, and the NFPA Production Services team who pitched in to highball the job through when it really counted. Other members of the production team who deserve mention are: Joyce Weston, Interior Design; Cameron, Inc., Cover and dust jacket design; and Peggy Holloway, Indexer.

Jim L. Linville
Product Manager

Fire Alarm
Signaling Systems

Second Edition

Fire Signatures

INTRODUCTION

Proper design of a fire alarm signaling system requires an understanding of fire itself—its byproducts, or "fire signatures"; its effects; how it can be detected; and the importance of installing a fire alarm signaling system. This chapter describes specific fire signatures and their measurement; effects of these signatures on humans; and how fire signatures reach fire detectors.

FIRE SIGNATURE FUNDAMENTALS

From the moment of its initiation, fire produces a variety of changes in the surrounding environment. Any of these changes in the ambient condition is referred to as a "fire signature" and can be monitored by a detection system. The production of smoke, for example, will result in a decrease in visibility that can be detected. To be useful, however, a fire signature should generate a measurable change in some ambient condition, and the magnitude of that change (the "signal") must be greater than the normal background variations (the "noise") for the condition. All other factors, e.g., hardware costs, and detection time being equal, the preferred fire signature will be that which can generate the highest signal-to-noise ratio in the earliest period of fire development. Fuel-specific signatures, such as the release of hydrogen chloride (HCl) from polyvinyl chloride (PVC) combustion, may be helpful in detectors for specialized applications but are of little use for general-purpose functions. Individual fire signatures are discussed below. The operational principles of the signatures and their associated detection mechanisms are discussed in a later section of this chapter.

Aerosol (Smoke) Signatures

The process of combustion releases very large numbers of solid and liquid particles into the atmosphere. The size of the particles ranges from 5×10^{-4} micrometers to 10 micrometers. These particles suspended in air are called aerosols and, when produced by fire, usually called smoke.

NFPA 72, *National Fire Alarm Code*, defines smoke as "the totality of the airborne visible or invisible particles of combustion." In NFPA 90A, *Standard for the Installation of Air Conditioning and Ventilating Systems*, the definition of smoke also includes "gases." According to one researcher[1]

> Smoke usually refers to a gaseous disperse system consisting of particles of low vapor pressure produced as a result of incomplete combustion processes. It takes the form of solid carbon particles and minute droplets of high molecular weight carbonaceous matter destructively distilled due to the heat of combustion.

Smoke particulate consists of liquid and solid particles covering a spectrum of particle sizes, depending upon the combustible materials and age of the smoke. As smoke ages and cools, the particles grow in size. This cooling effect relates to the stratification of smoke during pyrolysis, smouldering fires, and as the heat release rate from fire plumes weakens.

The essential feature of smoke is its instability. Under the influence of a lively Brownian motion, the particles in a cloud collide with one another and agglomerate. This process goes on continuously until the number of particles has been considerably diminished and the average size largely increased. The mass concentration of relatively stable smokes is usually low and, in the majority of cases, below a gram per cubic meter. High concentration leads to rapid coagulation and loss by sedimentation.

The aerosols resulting from a fire can be classified into two different fire signatures, depending on their ability to scatter light. Particles smaller than 0.3 micrometers in size are usually called *invisible*, since they do not scatter visible light efficiently. Particles larger than 0.3 micrometers do scatter visible light, and they are usually called *visible*.

"Visible" can be a relative term. Objects become visible when they reflect light energy from their surfaces into our eyes. This reflection is a function of the optical characteristics of the surface (color, reflectance). For small particles the ratio of the wavelength of the light relative to the particle diameter is also important—this ratio being α (λ/d). For $\alpha < 0.1$, particles will no longer scatter light sufficiently for it to be seen. This means that, given the range of light wavelengths that can be seen, particles below about 0.3 μm are invisible in visible light. Therefore, listed smoke-sensing fire detectors are tested only for response to smoke particles, and not to gases present in the smoke cloud.

When first introducing their fire detectors some manufacturers described the detectors as sensing products of combustion. Underwriters Laboratories Inc. (UL) listed both ionization and resistance bridge detectors as "combustion products detectors" in the early 1960s. After a UL research program was conducted in early 1968, however, these two types, along with photoelectric detectors, were categorized as smoke detectors, because all three are tested for response to smoke particles only.

The term "products of combustion" can be ambiguous in that aerosols, heat, light energy, and gases are, in fact, all products of the combustion process. Heating of materials during the pre-ignition stage of a fire produces submicron particles ranging in size from 5×10^{-4} to 1×10^{-3} micrometers. These particles are generated at temperatures well below ignition temperatures. The temperature at which submicrometer-sized particles are generated from materials is defined as the thermal particulate point. (See Table 1-1.) These invisible aerosols are generated in very large quantities. A 0.098-ounce (2.8 g) sample of bond paper burned in air in a 7.5 sq ft (2.28 m^2) room with an 8.9-ft (2.71-m) ceiling produced a maximum particle concentration of 3.6×10^{10} particles/cu ft (1.3×10^6/m^3).

As heating of a material progresses toward the ignition temperature, the concentration of invisible aerosols increases to the point where larger particles are formed by agglomeration. As this process continues, the particle size distribution becomes log normal, with the most frequent particle sizes in the 0.1 and 1.0 micrometer range. Particles less than 0.1 micrometer disappear either by agglomeration or evaporation, and the particles greater than 1.0 micrometer are lost through the process of sedimentation, following Stokes' Law. Aerosols in this latter size range are remarkably stable and contain both the visible and invisible particles. The production of visible aerosols can occur prior to ignition and is usually initiated at temperatures several hundred degrees higher than the thermal particulate point.

TABLE 1-1. Thermal Particulate Point in Air

Material	Temperature °F	°C
Bakelite™	380	191
PVC Insulation	290	142
Acrylan Carpeting	340	169
Wool	360	180
Paper	500	257
Pine Board	320	158
Polystyrene	710	373
Polyethylene	410	208
Motor Oil (SAE30)	310	153

Test results from smoldering and flaming combustion of various materials indicate that, generally, smoldering fires produce more large particles than flaming fires.[2] It is important to note that the maximum relative particle concentration for both fire types appears to be in the range of particle sizes smaller than 0.3 micrometer. Particle size seems to affect the optimum detection method. The condensation nuclei-(cloud-chamber) type detector is best suited for the smaller invisible aerosol signatures, and the ion chamber type for the larger. Photoelectric units containing a light source with a major spectral component in the near-ultraviolet and blue-green wavelengths, plus a suitable photocell, should also respond to the larger invisible aerosols, since the best scattering of energy occurs when the particle diameter approaches the wavelength of the incident radiation.

Energy Release Signatures

In all stages, fire releases energy into the surrounding environment. This energy release produces several useful fire signatures.

IR and UV Radiant Energy: Infrared (IR) and ultraviolet (UV) signatures are usually the earliest energy signatures that can be detected from a fire. With the exception of acetylene and other highly unsaturated hydrocarbons, the infrared emissions from hydrocarbon particles are particularly strong when particles are in the 4.4-micrometer region [due to carbon dioxide (CO_2)] and in the 2.7-micrometer region [due to water vapor (H_2O)]. These emissions account for nearly all the emitted energy. Since the infrared component of sunlight (which is a potential "noise," or false signal, source) is reduced in these regions by absorption due to atmospheric carbon dioxide and water, a high signal-to-noise ratio is obtained. The carbon dioxide-water radiation signature can be used effectively for detection, but there is the possibility of noise from man-made infrared sources. The modulation of the energy output level due to flame flicker is another signature that affects the infrared signal from a flame. This flicker is characteristic of flame and has a frequency range of 5 to 30 Hz.

The total infrared signature has been used in the detection of both smoldering and flaming fires. The disadvantage of the infrared signature is that it has a wide range of noise levels from solar and man-made sources. Detectors that use both the carbon dioxide (CO_2)-water vapor (H_2O) and flicker signatures have an excellent signal-to-noise ratio.

The ultraviolet fire signatures appear in flames as emissions from hydroxal (OH), carbon dioxide (CO_2), and carbon monoxide (CO), ranging in size from about 0.27 to 0.29 micrometer. Although the signal-to-noise ratio of the UV signatures is less than that of the IR signatures, UV signatures have been effectively used for flame detection. They are especially effective where magnesium or its alloys may be involved, because of the strong signal that burning magnesium causes in the 0.28 micrometer region.

Thermal Energy: Convected thermal energy from a fire causes an increase in the air temperature of the surrounding environment. The time required for release of sufficient energy to produce a significant convected energy signal varies from less than one minute for rapidly developing fires to hours for slowly developing/smoldering deep-seated fires. In contrast to aerosol and radiant energy signatures, convected thermal energy signatures often appear well after life-threatening conditions have been reached from excessive aerosol and/or toxic gas concentrations.

Gas Signatures

During a fire, many changes occur in the gas content of the atmosphere. These changes can be found both in the area of fire origin and in areas far removed from the fire. These changes in atmospheric gas content are mostly due to the addition of gases that are not normally present. These changes are termed "evolved gas signatures."

Another gas signature that occurs during a fire is the reduction of oxygen content; this is termed the "oxygen depletion signature."

Many gases evolve during a fire, e.g., water vapor (H_2O), carbon monoxide (CO), carbon dioxide (CO_2), hydrogen chloride (HCl), hydrogen cyanide (HCN), hydrogen fluoride (HF), hydrogen sulfide (H_2S), ammonia (NH_3), and nitrogen oxides. Most of these evolved gas signatures are fuel specific, and, since they are not associated with a sufficiently large number of fuels, cannot be used for general-purpose detection.

The rate at which any gas is produced in a fire is a relatively constant function of the mass burning rate of the fuel. Referred to as the *yield fraction*, this is the fraction of the burned mass that results in gas. Such yield fractions are a function of the specific fuel, and for oxygenated species depend on the amount of combustion air available.

Carbon monoxide (CO) is present in nearly all fires. Tests have shown a potential use for a carbon monoxide signature in fire detection. The rate of carbon monoxide production varies considerably with the type of fuel, the amount of ventilation, and whether the fire is of the smoldering or open-flaming type. Yield fractions for CO are typically a few percent in fully ventilated fires but can increase by a factor of 30 to 50 at flashover. Yield fractions in smoldering fires can be just as high, but, due to low burning rates, the quantity of CO produced is low.

When CO has been examined as a fire detection signature, it is generally found that detectable levels of CO (75-100 ppm) occur after detectable levels of particulates. Researchers are currently studying the use of CO in conjunction with particulates as the basis of multi-mode sensors that can descriminate unwanted fires from the more common false-alarm sources, e.g., cooking and steam from showers, since they lack the CO component.

FATALITY POTENTIAL FOR FIRE SIGNATURES

Fire signatures create a hostile and often fatal environment. Exposure to signatures produces psychological and physiological effects, both individually and in combination, in humans.

Effects of Aerosols

Aerosols derived from a fire produce several effects on humans. Much of the aerosol consists of carbon particles upon which may be absorbed a number of irritant compounds, such as organic acids, aldehydes, or hydrogen chloride. Studies with dogs exposed to kerosene aerosol consisting largely of carbon soot produced no noticeable respiratory damage. Most of the soot was expelled, or coughed out, by the dogs within 24 hours. This study further indicated that the aldehydes in wood aerosol were the significant factor in causing respiratory tract swelling and irritation.

Inhalation of these products makes breathing difficult and may be accompanied by severe coughing, a burning sensation in the chest and throat, and irritation of the eyes. In addition to these physiological effects, absorption and scattering of light from visible aerosols can reduce visibility to near zero in some cases. Physiological and environmental changes, combined with fear, anxiety, and loss of orientation in a fire situation, often result in panic behavior. Persons who panic in a fire can act irrationally and may become lost or trapped in dead-end corridors or closets. In extreme cases, persons become "paralyzed" and take no lifesaving action at all.

Visible aerosol concentration is measured in terms of percent reduction of light per foot (meter) of travel path. In a Japanese study it was found that persons familiar with their surroundings can tolerate a maximum of 9.2 percent per ft (0.13 OD/m) reduction of light, while those unfamiliar with their surroundings usually only have a 3 percent per ft (0.04 OD/m) tolerance.

Effects of Heat

Heat from fire primarily exposes humans to the potential of burn injury. Experiments with animals have shown that death occurs after 2 minutes of exposure at 212°F (100°C) and at 30 minutes when exposed to 140°F (60°C). Although other tests[3] have shown that the ability to voluntarily endure heat can vary from 2 minutes at 300°F (149°C) to 45 seconds at 740°F (393°C), it is likely that these are extreme levels of tolerance. In addition to surface burns, inhalation of heated gases and aerosols often produces such thermal damage as sloughing of the trachea lining and hemorrhaging in the respiratory track. Exposure to elevated temperatures can also cause shock, since blood pools at the body surface as the body attempts to cool itself.[4]

Effects of Toxic Gases

Carbon monoxide, carbon dioxide, and other gases can be generated in sufficient quantity in a fire to produce toxic reactions in humans.

Carbon Monoxide: Carbon monoxide (CO) is present in nearly all hostile fire situations and is an important factor in loss of life from fire. Clinical examination of autopsy reports of fire victims in New York City showed that carbon monoxide poisoning, rather than respiratory tract damage, was the significant factor in death with victims having post-burn survival times of less than 12 hours.[5] The usual threshold limit value for exposure to carbon monoxide for an 8-hour working day is 50 parts per million (ppm). In fires with conditions of restricted ventilation, carbon monoxide concentration as high as 138,000 ppm have been recorded.

Carbon monoxide reduces the oxygen-carrying ability of the blood by combining with hemoglobin to form carboxyhemoglobin. The various effects of carbon monoxide exposure (in order of occurrence) include: headache, dizziness, dimness of vision, nausea, increased pulse and breathing rates, confusion and loss of orientation, unconsciousness, reduced pulse and breathing rates, convulsions, and, ultimately, death. Exposure to even moderate levels of carbon monoxide can severely restrict a person's ability to function in a fire situation, especially when combined with the adverse effects of aerosols and elevated temperatures. (See Table 1-2.) A 1973 study of 106 fire fatalities found that 50 percent of those involved carbon monoxide poisoning as the cause of death. An additional 30 percent of the fatalities were traced to carbon monoxide poisoning that complicated preexisting effects of heart disease or alcohol use.[6]

Carbon Dioxide: Carbon dioxide (CO_2) is one of the products of complete combustion and is commonly associated with both flaming and

TABLE 1-2. Effects of Carbon Monoxide Exposure

Percent Concentration	ppm	Exposure Time	Effects
0.02	200	2–3 hrs	Mild Headache
0.08	800	45 min	Mild Headache
		2 hrs	Possible Death
0.32	3,200	10–15 min	Dizziness
		30 min	Death
0.69	6,900	1–2 min	Dizziness
		10–15 min	Death
1.28	12,800	2–3 breaths	Unconsciousness
		1–3 min	Death

smoldering fires in amounts exceeding the threshold limit value of 5,000 ppm. Carbon dioxide stimulates the breathing rate; it may also act as an asphyxiant by displacing oxygen. The early symptoms of carbon dioxide intoxication are dizziness and shortness of breath followed by mental excitement and often irrational behavior. Concentrations of 9 percent (90,000 ppm) CO_2 in the atmosphere can be fatal in 4 hours due to carbon dioxide-hemoglobin complexing. Although carbon dioxide concentrations up to 6.75 percent have been recorded in bedding test fires, even low concentrations of carbon dioxide will raise the breathing rate and increase the intake of other toxic substances. A 5 percent concentration of CO_2 is used to increase the breathing rate of smoke inhalation victims, speeding the process of elimination of CO bound in the blood. (See Table 1-3.)

TABLE 1-3. Effects of Carbon Dioxide Exposure

Percent Concentration	ppm	Effects
0.5	5,000	Increased depth of breathing
3.0	30,000	Breathing rate doubles
5.0	50,000	300% increase in breathing rate
10.0	100,000	Possible death, even with sufficient atmospheric oxygen

Hydrogen Chloride: Hydrogen chloride (HCl) is not found in cellulosic-based fires but is often associated with the burning of many common plastics, such as vinyls. The threshold limit value is 5 ppm, and a noticeable irritation of the mucous membranes occurs at 35 ppm. Concentrations of 1,000 to 2,000 ppm produce severe reactions that may be fatal.

Other Gases: Many other gases evolve from burning materials in small but potentially dangerous quantities. Tests of both flaming and smoldering combustion of materials used in aircraft interior finishing produced measurable quantities of toxic gases.[7] In the tests, burning polyamides and modacrylics produced hydrogen cyanide. Hydrogen cyanide is also produced from burning hair, leather, wool, acrylics, polyurethane foams, and some dyes, as well as some synthetic fabric finishes.

TABLE 1-4. Effects of Oxygen Depletion*

Percent Concentration	Exposure Time	Effects
21–17	Indefinite	Respiration volume decreases, loss of coordination and difficulty in thinking
17–14	2 hrs	Rapid pulse and dizziness
14–11	30 min	Nausea, vomiting, and paralysis
9	5 min	Unconsciousness
6	1–2 min	Death within a few minutes

*Approximate figures.

Hydrogen sulfide (H_2S) was produced from chloroprene. Polysulfone plastic was found to produce sulfur dioxide (SO_2) which can also be produced from burning rubber and wood. Hydrogen fluoride, nitrogen oxides, and ammonia were also detected in the tests.

This is only a small percentage of the chemical species present in fire gases. Although usually present only in small sub-lethal quantities, these gases can produce additive and synergistic effects within the total system of fire products. These effects will be discussed below.

Effects of Oxygen Depletion

Although oxygen depletion is usually confined to the immediate area of a fire, its effects can be felt in remote areas. Oxygen depletion occurs both in a large open burning fire or, if the building is tightly sealed, smoldering fires of long duration. The physiological effects of oxygen depletion are shown in Table 1-4.

Additive and Synergistic Effects of Toxic Materials

Additive and possible synergistic effects must be considered when dealing with mixtures of toxic materials. In a decade of research, combustion toxicologists have determined that most observed toxic effects in fires are the result of exposure to the short list of toxic species discussed above. The interrelationship of these gases is expressed as the *N-Gas Model*:

$$FED = \frac{m[CO]}{[CO_2]-b} + \frac{[HCN]}{LC_{50}(HCN)} + \frac{21-[O_2]}{21-LC_{50}(O_2)} + \frac{[HCl]}{LC_{50}(HCl)} + \frac{[HBr]}{LC_{50}(HBr)}$$

This equation is valid for 30-minute exposures to a mixed gas stream and predicts fatalities during exposure or for 14 days post-exposure when the fractional effective dose (FED) > 1. [FED = dose received at time (t)/effective dose to cause incapacitation or death] Note that CO_2 appears only as a mechanism to potentiate the CO toxicity, but is not toxic itself.

Combining the effects of increased breathing rates associated with elevated carbon dioxide and carbon monoxide and decreased oxygen, plus the irritant effects of fire-generated aerosols and the possible synergistic effects of small amounts of other toxic gases, it is clear that persons exposed to these conditions can quickly be rendered helpless. Thus, quick-response detection devices are critical to life safety in fire situations.

Numerous tests have demonstrated that smoke detectors are the most practical fire detectors for use as an early warning device. One conclusion of the second Indiana Dunes Tests[8] program stated, in part,

> Supporting the first-year results, the fixed-temperature heat detectors rated for 50-ft (15-m) spacing [135°F (57°C)] used in this test series, in the room of fire origin, provided little life-saving potential. These detectors failed to respond to a majority of the fires and when they did respond they were considerably slower than smoke detectors located remotely from the fire.

Similar conclusions were drawn in ten studies conducted over twenty years, and are summarized in a recent review article.[9]

TRANSPORT OF FIRE SIGNATURES

A fire detector is a device with a preset sensitivity level. When the area (or conditions) immediately surrounding the detector exceed those preset parameters, a signal is produced. First, fire detectors are designed to respond to matter or energy that is usually but not exclusively associated with combustion. False alarms occur when identical or similar matter or energy is produced within the protected space by controlled combustion or noncombustion sources. Second, all fire detectors have a threshold value, or sensitivity, for sensed parameters that must be exceeded before the detector produces a signal. Third, the detector can only respond to the level of the sensed parameter present at its location—i.e., regardless of the amount of matter or energy being released by the fire, only that which actually reaches the detector can be sensed. The mechanisms associated with the transport of matter and energy from the fire to the detector are critical to detector response.

Fire detectors typically sense one of three parameters: (1) convected heat, (2) smoke particles, or (3) radiated energy of mass or energy released by combustion. The generation and transport of each parameter differs, and the differences should be understood for optimum detector use.

Fire is a generic term that refers to flaming combustion, smoldering combustion, or pyrolysis. Flaming combustion usually releases considerable quantities of convected heat, smoke particulate, and radiant energy. Smoldering, however, produces only particulate which has considerably different physical properties from particulate produced by flaming combustion. True smoldering only occurs in char-forming materials, since the thermal insulating nature of the char is necessary to trap the heat and maintain the glowing reaction zone. Pyrolysis is the thermal degradation process in non-char-forming materials which produces particulate similar to smoldering but without a glowing reaction zone.

Flaming combustion is easiest to detect because it produces all the signatures to which detectors are sensitive, and the rate of heat release from the combustion is generally sufficient to rapidly transport the particulate to the detector area.

Smoldering and pyrolysis, in contrast, are only adequately detected by smoke detectors. Delays in detection can occur because the limited quantity of heat liberated provides little buoyancy that can drive the particulate to the detector location. The smoke that is produced moves more slowly, is less concentrated in a stratified layer, and is much more subject to the influences of forced convection (heating, ventilating, and air conditioning) systems and other influences.

For detectors that respond to infrared or ultraviolet flame radiation, the energy transport process occurs at the speed of light. The only potential interference with this transport process is blockage of the signal by a physical object. Infrared radiation will reflect off wall surfaces so, even if the detector is not in the direct line of the fire, response can still occur. Response time in this case will be delayed because energy dissipates in the surface reflection process. Ultraviolet-type flame detectors, however, do not respond to reflected energy and must be within a direct line of the flame for response.

Losses are another important factor in the transport process. With thermal energy, the temperature of the fire gases is reduced by entrainment (mixing) of cool air into the fire plume and by convective heat transfer to surfaces. It is often not understood that, when the fire plume first reaches the ceiling and begins to spread outward, as much as 90 percent of the energy contained in the layer is lost to the ceiling. This is one reason why heat detectors are the slowest of fire detectors.

The principal mechanism by which the smoke concentration is reduced is dilution, as clean air is entrained into the plume. While surface losses do occur, the mass loss to surfaces is not significant in reducing the smoke concentration in the upper air layer.

Flame radiation, as with any radiated energy, decreases in intensity with the square of the distance between the radiating source and the target. Some flame detectors can be fitted with concentrators that reduce the effective field of vision and allow the detector to respond at longer distances to the same radiated energy level.

CONCLUSION

The characteristics of the various fire signatures, with their potential to cause serious fire injury or death, are important considerations in fire alarm system design. Design parameters for signal initiation, transmission, and processing; electrical supervision and installation; and power supply sources are based on fire signatures. Succeeding chapters of this text describe further how fire signatures affect each step of fire alarm signaling system design.

BIBLIOGRAPHY

NFPA Codes, Standards, Recommended Practices, and Manuals. (See the latest *NFPA Catalog* for availability of current editions of the following documents.)

NFPA 72, *National Fire Alarm Code*

NFPA 90A, *Standard for the Installation of Air Conditioning and Ventilating Systems.*

References Cited

1. Haessler, W., "Smoke Detection by Forward Light Scattering," *Fire Technology*, Vol. 1, No. 1, 1965, p. 43.
2. Scheidweiler, A., "New Research in Fire Detection Technology," International Symposium, 1968.

3. Dupont, H.B., "How Much Heat Can Firemen Endure?," *Fire Engineering*, Vol. 113, No. 2, 1960, pp. 122-214.

4. Zikria, B.A., *et al.*, "Respiratory Tract Damage in Burns: Pathophysiology and Therapy," *Annals of New York Academy Sciences*, Vol. 150, Art. 3, 1968, pp. 618-626.

5. Zikria, B.A., *et al.*, *Smoke and Carbon Monoxide Poisoning in Fire Victims*, American Association for the Surgery of Trauma, New York, 1971.

6. "Annual Summary Report: 1 July 72-30," June 1973, Johns Hopkins University Applied Physics Lab, Baltimore, MD.

7. Gross, D., *et al.*, *Smoke and Gases Produced by Burning Aircraft Interior Materials*, NBS Building Science Series 18, US National Bureau of Standards, Washington, DC, 1969.

8. Indiana Dunes Tests (2nd series)

9. Bukowski, R. W., "Studies Assess Performance of Residential Detectors," *NFPA Journal*, Vol. 87, No. 1, Jan./Feb. 1993, pp. 48-54.

Additional Reading

Los Angeles Fire Department, "Fire Detection Systems in Dwellings," *NFPA Quarterly*, Vol. 56, No. 3, 1963, pp. 201-246.

Pryor, A.J., *et al.*, *Hazards of Smoke and Toxic Gases Produced in Urban Fires*, Southwest Research Institute, San Antonio, TX, 1968.

Harpe, S. W., Waterman, T. E., and Christian, W. J., "Detector Sensitivity Siting Requirements for Dwellings, Phase 2," IIT Research Institute, Chicago, IL, 1977.

2

Fire Alarm Systems

INTRODUCTION

A fire alarm signaling system is a key element of any building's overall fire protection features. Properly designed, installed, operated, and maintained, a fire alarm system can help limit fire losses in buildings, regardless of occupancy. (NB: the terms "fire alarm system" and "protective signaling system" are used interchangeably throughout this text.) Also, since many of the fire deaths in the United States result from building fires, the use of detection and alarm systems in buildings can reduce the loss of life from fire. This chapter describes the operational characteristics of the various types of protective signaling systems currently in use.

DESCRIPTION OF A FIRE ALARM SYSTEM

A fire alarm system is a system that is primarily intended to indicate and warn of abnormal conditions, summon appropriate aid, and control occupant facilities to enhance protection of life.

A fire alarm system either automatically or manually links the sensing of a fire condition and the notification of people within and outside of the building that action should be taken to respond to the fire condition. The system might also be designed to initiate that response by initiating operation of a fixed extinguishing system, such as a pre-action or deluge sprinkler system; a carbon dioxide, halon, or foam system; etc.

The basic components of a fire alarm system are one or more alarm initiating device circuits to which automatic fire detectors, manual fire

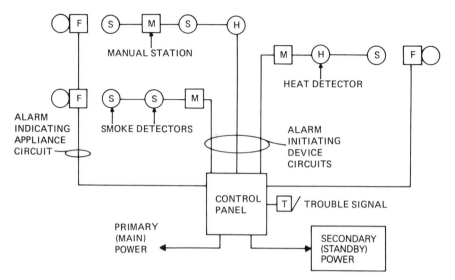

M,S,H = ALARM INITIATING DEVICES
F = ALARM INDICATING APPLIANCE

Figure 2-1. Basic components of a fire alarm system.

alarm boxes, waterflow alarm devices, or other alarm initiating devices are connected. These circuits carry alarm indication to a control panel, which is powered by a primary (main) power supply and a secondary (standby) power supply. Components of a fire alarm system also include one or more alarm indicating appliance circuits to which alarm signals, such as bells, horns, speakers, etc., are connected, or to which an off-premises alarm is connected, or both; and a trouble signal. (See Figure 2-1.)

TYPES OF SYSTEMS

Fire alarm systems are classified according to the type of functions they are expected to perform. Their installation, maintenance, and use are specified in NFPA 72, *National Fire Alarm Code*. (See Table 2-1.)

Power Supply Requirements for Fire Alarm Systems

Regardless of the type of alarm system, NFPA 72, *National Fire Alarm Code*, specifies similar requirements for protective system power supply.

Primary (Main) Power Supply: The primary (main) power supply may be provided by commercial light and power supply or an engine-driven

TABLE 2-1. Types of Fire Alarm Signaling Systems

Type	Description	Comments
1. Local protective signaling system.	All alarm system operating in the protected premises, responsive to the operation of a manual fire alarm box, waterflow in a sprinkler system, or detection of a fire by a smoke or heat detecting system.	The main purpose of this system is to provide an evacuation alarm for the occupants of the building. Someone must always be present to transmit the alarm to fire authorities. See NFPA 72, *National Fire Alarm Code.*
2. Auxiliary protective signaling system.	An alarm system utilizing a standard municipal coded fire alarm box to transmit a fire alarm from a protected property to municipal fire headquarters. These alarms are received on the same municipal equipment and are carried over the same transmission lines as are used to connect fire alarm boxes located on streets. Operation is initiated by the local fire detection and alarm system installed at the protected property.	Direct means of summoning help from municipal fire department. Some communities will accept this type of system and others will not. See NFPA 72, *National Fire Alarm Code.* Trouble signal may register in a separate attended location.
3. Remote station protective signaling system.	An alarm system connecting protected premises over leased telephone lines to a remote station, such as a fire station or a police station. Includes separate receiver for individual functions being monitored, such as fire alarm signal, or sprinkler waterflow alarm.	Requires leased lines into each premise. See NFPA 72, *National Fire Alarm Code.*
4. Central station protective signaling system.	An alarm system connecting protected premises to a privately owned central station whose function is to monitor the connecting lines constantly and record any indication of fire, supervisory or other trouble signals from the protected premises. When a signal is received, the central station will take such action as is required, such as informing the municipal fire department of a fire or notifying the police department of intrusion.	Flexible system. Can handle many types of alarms, including trouble within system at protected premises. See NFPA 72, *National Fire Alarm Code.*
5. Proprietary protective signaling system.	An alarm system that serves contiguous or noncontiguous properties under one ownership from a central supervising station at the protected property. Similar to a central station system but owned by the protected property.	Requires 24-hr manning of central supervising station on the premises. See NFPA 72, *National Fire Alarm Code.*
6. Emergency voice/alarm communication system.	Provides for the inclusion of emergency voice/alarm communications in any of the systems listed above.	Provides dedicated facilities for the transmission of information to occupants of the building (including fire department personnel). See NFPA 72, *National Fire Alarm Code.*

generator. The commercial supply can consist of one phase of three-wire supply, or two-wire supply. The connection to light and power supply must be on a dedicated branch circuit, with the disconnecting means marked "fire alarm circuit control," and accessible only to authorized personnel.

An alternate primary power supply is an engine-driven generator consisting of a single engine-driven generator with either (1) a trained operator on duty 24 hours a day or an engine-driven generator and a storage battery with 1-hour capacity, or (2) multiple engine-driven generators one of which must be arranged for automatic starting.

The primary power supply must be supervised and its failure indicated by a distinctive trouble signal.

Secondary (Standby) Power Supply: The secondary (standby) power supply must provide power to the complete system within 30 seconds in event of primary supply failure or whenever the primary power supply drops below 80 percent of rated value. The secondary supply must have the capacity requirements as listed in Table 2-2.

TABLE 2-2. Secondary Power Supply Requirements

System Type	Maximum Normal Load	Alarm Load
Local System	24 hrs	5 min
Auxiliary System	60 hrs	5 min
Remote Station System	60 hrs	5 min
Proprietary System	24 hrs of normal traffic	
Emergency Voice/Alarm Communication System	24 hrs	5 min

Acceptable sources of standby power are: a storage battery and charger; or an engine-driven generator and 4-hour capacity storage battery; or multiple engine-driven generators, one of which must be arranged for automatic starting.

Trouble Signal Power Supply: A trouble signal power supply for fire alarm systems operates a trouble signal when the primary (main) power supply fails. The trouble signal power supply must be independent of the primary supply (but can be the secondary power supply) and cannot use dry cells.

Local Protective Signaling Systems

Local protective signaling systems, as specified in NFPA 72, *National Fire Alarm Code*, are supervised systems that provide fire alarm and supervisory signals within a facility and produce a signal at the facility only.

The main purpose of a local protective signaling system is to sound local alarm signals for evacuation of the protected building. A system could be limited to the basic features of a fire alarm system (Figure 2-1), or may have other features such as an annunciator, fire door control, return elevator, or voice communication within the protected property. (See Figure 2-2.)

Figure 2-2. Remote annunciator.

A local fire alarm system must be powered by a primary (main) power supply, a secondary (standby) power supply, and a trouble signal power supply. The secondary supply must be capable of operating the system for a 24-hour period under normal load and subsequently operate the system in alarm for an additional 5 minutes. This power supply covers a normal day of occupancy and then a 5-minute alarm period, if necessary, to evacuate the building. (See Figure 2-1.)

In a local protective system, the alarm is not automatically relayed to a fire department; instead, someone must notify the fire department when the alarm sounds. If the building is unoccupied at the time of the alarm, fire department response would depend on a neighbor or passerby hearing the alarm and notifying the department. (See Figure 2-3.)

IONIZATION   STROBE HORN
SMOKE DETECTOR

LOCAL FIRE
ALARM PANEL

Figure 2-3.   Typical equipment used in a local protective signaling system.

A local protective system may carry out several functions in a building, e.g., automatic fire detection; manual fire alarm; guard's tour supervisory service; sprinkler system waterflow alarm and supervisory signal service; combination systems (e.g., multipurpose electrical system, burglar alarm, music distribution, and paging systems—whether the voice or coded type); smoke control; and control of fire doors, elevator return, or air-handling systems when actuated by the fire alarm system. A local system connected to wireless smoke detectors is shown in Figure 2-4.

A key element of local protective signaling systems is that all devices, appliances, and equipment in the system must be listed for its intended use. Both Underwriters Laboratories Inc. and Factory Mutual Research

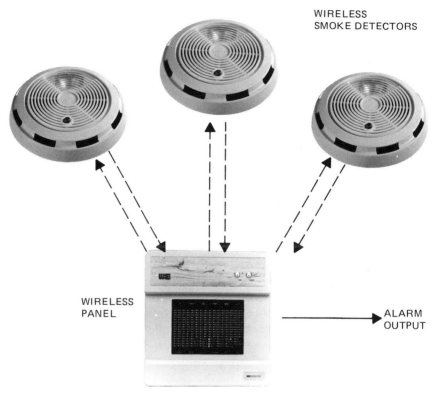

WIRELESS
SMOKE DETECTORS

WIRELESS
PANEL

ALARM
OUTPUT

Figure 2-4. Typical equipment used in a wireless local system.

Corporation have listed equipment for this service, and other testing laboratories may be acceptable to the authority having jurisdiction. The listed control units are normally tested for ordinary indoor use only. The installation of control units in an outdoor environment is not recommended.

The local protective system may perform automatic control functions, such as release of smoke doors and smoke dampers or deenergizing computer equipment and heating, ventilating, and air conditioning (HVAC) systems. Automatic controls should not impede building lighting, as this may hamper building evacuation and life safety.

Alarm signals can be coded or noncoded. Coded alarm signals consist of at least three complete rounds of the number transmitted. Coded supervisory signals usually have a single round of the number transmitted.

Use of a standard fire alarm evacuation signal is now being phased in for building evacuation. The signal is a uniform Code 3 temporal pattern, using a sound keyed $1/2$ to 1 second "ON," $1/2$ second "OFF," $1/2$ to

1 second "ON," $\frac{1}{2}$ second "OFF," $\frac{1}{2}$ to 1 second "ON," and $2\frac{1}{2}$ seconds "OFF." The signal should be repeated for at least three minutes. This signal is for evacuation purposes, and its use should be restricted to occupant evacuation. This fire alarm evacuation signal pattern is normally *not* used in health-care facilities and penal institutions unless complete evacuation is necessary.

Power supply for a local protective signaling system is normally supplied from the commercial power and light service because of its high reliability. As for all alarm systems, a dedicated branch circuit that is mechanically protected should be connected to the power and light service of the facility and be marked "fire alarm circuit control." Engine-driven generators can also provide this power to the system.

Standby, or secondary, power supplies are now required in local systems. In addition to the main power supply, a trouble power supply must be provided to indicate failure of the primary power supply or failure of an electrically supervised circuit. Dry-cell batteries are not permitted for use as the trouble power supply due to their unreliability over time and their lack of recharging capability. The secondary power supply can also be used as the trouble power supply. Secondary power supplies should be capable of providing power under normal system load for 24 hours and then operate the system in alarm condition for at least 5 minutes. Automatic transfer between the main and secondary power supplies should be completed within 30 seconds.

Auxiliary Protective Signaling Systems

Auxiliary protective signaling systems are supervised systems in a protected facility that transmit alarm signals via a municipal fire alarm system. The system is intended to automatically call or notify a fire department following activation of an automatic or manual fire alarm. (An auxiliary system may or may not have an evacuation alarm). The municipal system then transmits an alarm to the municipal communication center.

The alarms from an auxiliary protective system and the municipal fire alarm boxes located in the community both transmit their signals with the same equipment and communication method. With an auxiliary system, the owner of the protected facility is responsible for system operation, maintenance, and testing, while the municipality is responsible for its leased or owned system.

An auxiliary protective signaling system circuit connects the alarm initiating devices to a municipal fire alarm system. The connection is made either through a nearby transmitter or a master fire alarm box or through a dedicated telephone line run directly to the municipal connection center switchboard. The signal received by the fire department is identical to that received by the department upon manual activation of the municipal fire alarm box. Because fire department personnel know which municipal boxes are auxiliarized, responding fire fighters can check for an alarm originating within the protected facility. (See Figure 2-5.)

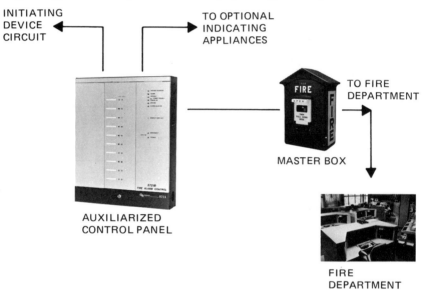

Figure 2-5. Connections in an auxiliary signaling system.

There are three types of auxiliary systems recognized in NFPA 72, *National Fire Alarm Code*: (1) the local energy type, (2) shunt type, and (3) parallel telephone type.

The local energy-type system (Figure 2-6 and Figure 2-7) is defined in NFPA 72, *National Fire Alarm Code*, as electrically isolated from the municipal alarm system and has its own power supply. Tripping of the transmitting device does not depend on the current in the municipal system. In a wired circuit, receipt of the alarm by the municipal communication center (if the municipal circuit is accidentally opened) depends on the design of the transmitting device and the associated municipal communication center equipment, i.e., whether or not the municipal system is designed to receive alarms through manual or automatic ground operational facilities. In a radio box-type system, receipt

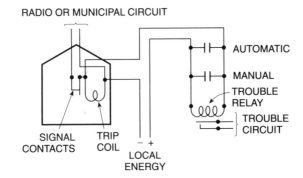

Figure 2-6. Local energy-type auxiliary alarm system.

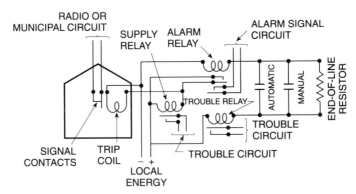

Figure 2-7. Local energy-type auxiliary alarm system with local alarm.

of the alarm by the municipal communication center depends upon the proper operation of the radio transmitting and receiving equipment.

The shunt-type system, as described in NFPA 72, *National Fire Alarm Code*, is electrically connected to and is an integral part of the municipal alarm system. (See Figure 2-8 and Figure 2-9.) A ground fault on the auxiliary circuit is a fault on the municipal circuit, and an accidental opening of the auxiliary circuit will send a needless (or false) alarm to the municipal communication center. An open circuit in the transmitting device trip coil will not be indicated either at the protected property or at the municipal communication center; also, if an alarm initiating device is operated, the alarm will not be transmitted but an open circuit indication will be given at the municipal communication center. If a municipal circuit is open when a connected shunt-type system is operated, the transmitting device will not trip until the municipal circuit returns to normal, at which time the alarm will be transmitted.

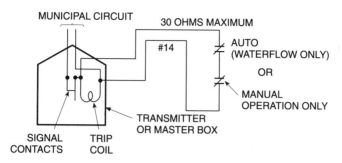

Figure 2-8. Shunt-type auxiliary alarm system.

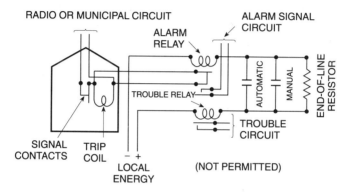

Figure 2-9. Shunt-type master box and local energy-type auxiliary alarm system.

NFPA 72, *National Fire Alarm Code*, specifies that shunt-type systems are to be used for manual fire alarm and waterflow alarm service only. Automatic fire detection is not permitted on shunt-type systems. This system is noncoded with respect to any remote electrical tripping or actuating devices.

Additional design parameters for shunt-type systems are:

1. All conductors of the auxiliary circuit must be in conduit or electrical metallic tubing in accordance with NFPA 70, *National Electrical Code®*, Articles 346 and 348.
2. Both sides (conductors) of the auxiliary circuit must be in the same conduit.
3. Maximum permitted shunt loop length is 750 ft (228 m) in conduit.
4. Smallest conductor for the auxiliary circuit is 14 AWG (NFPA 70, *National Electrical Code*, Article 310).

5. A local fire alarm system auxiliarized by the addition of a relay coil energized by a local power supply and whose normally closed contacts trip a shunt-type master box is not permitted. (See Figure 2-9.)

6. Where a municipal transmitter or master box is located within a private premise, it must be installed in accordance with NFPA 1221, *Standard for the Installation, Maintenance, and Use of Public Fire Service Communication Systems*. (See Chapter 15.)

A parallel telephone-type auxiliary alarm system is described in NFPA 72, *National Fire Alarm Code*, as a system in which alarms are transmitted over a circuit directly connected to the annunciating switchboard at the municipal communication center, and terminated at the protected property by an end-of-line resistor or equivalent. Such auxiliary systems are for connection to municipal fire alarm systems of the type in which each municipal alarm box annunciates at the municipal communication center switchboard by individual circuit. (See Figure 2-10.)

There are major differences among these three types of auxiliary systems:

1. The shunt-type system is a series closed loop for the alarm initiating circuit. The local energy-type and parallel telephone-type are open contact parallel circuits for the alarm initiating circuit.

2. An accidental open circuit on a shunt-type system activates a fire alarm. An accidental open circuit on a local energy-type or parallel telephone-type system activates a trouble alarm.

3. A single ground fault on the alarm initiating circuit of either the shunt- or parallel telephone-type system extends into the municipal box wired circuit or switchboard.

4. A single ground fault on a local energy-type system that prevents normal operation on the alarm initiating circuit causes a trouble signal.

Some of the basic requirements for auxiliary protective signaling systems include:

1. One municipal transmitter or master box that must serve a maximum 100,000 sq ft (9290 m²) fire area,

2. Means to manually initiate an alarm signal when automatic fire detectors and alarm service, and waterflow alarm services are provided, and

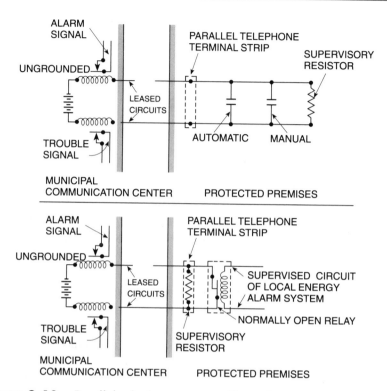

Figure 2-10. Parallel telephone-type auxiliary alarm system.

3. An annunciator, located near the transmitting device.

Specific power sources are required by NFPA 72, *National Fire Alarm Code*, for these auxiliary systems: the municipal fire alarm system for the shunt-type system, a primary (main) *and* secondary (standby) power supply for the local energy-type system, and the municipal communication center as the power supply for the parallel telephone-type auxiliary system. The standby power supply for all types of auxiliary systems is required to operate the system for 60 hours of normal operation, followed by 5 minutes of alarm. Since these systems are installed primarily for building protection, the 60-hour span is necessary to carry the system over a weekend period.

The auxiliary alarm systems do not require the use of audible alarm signals within the facility. If fire alarm evacuation signals are needed for the protected property, the system should also comply with the provisions of NFPA 72, *National Fire Alarm Code*, Chapter 6.

A design engineer should work closely with the municipal fire department communication center to ensure the auxiliary protective signaling system is compatible with the municipality's system. This compatibility is critical to efficient auxiliary system operation.

Remote Station Protective Signaling Systems

Remote station protective signaling systems are supervised systems in one or more protected facilities that transmit alarm, trouble, or supervisory signals to the remote station through a direct connection (usually a leased telephone line) from each protected property as described in NFPA 72, *National Fire Alarm Code*. They are attended by trained personnel 24 hours a day, usually at a fire station. Properties under the same or different ownership can be protected by a remote station. (See Figure 2-11 and Figure 2-12.)

When permitted by the authority having jurisdiction, the remote station may be in a location other than a fire station where personnel are on duty 24 hours a day and are trained to receive the alarm signal and immediately retransmit it to the fire department. System trouble signals usually are not automatically transmitted to the remote station. (A remote station system may or may not have an evacuation alarm.)

A remote station differs from a central station system primarily in that central station personnel are principally responsible for furnishing and maintaining supervised signaling service.

Acceptable methods of transmitting alarm signals from a remote station to the fire department (assuming the fire department is not the remote station) include:

1. A dedicated circuit independent of any switching network using voice communication or coded signals;
2. A one-way (outgoing only) telephone using a commercial dial network, or
3. A radio system with fire department frequency (where permitted).

The dedicated independent circuit is the preferred transmission method. Upon receipt of a supervisory or trouble signal, the remote station operator on duty is responsible for notifying the owner or owner's representative immediately.

NFPA 72, *National Fire Alarm Code*, requires an independent secondary (standby) power supply for the remote station control unit and, if needed, at the protected premises. The secondary supply power source usually can be any permitted for a fire alarm system, but it must

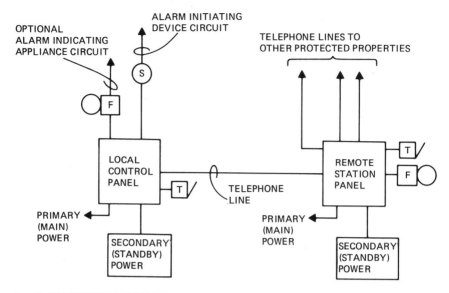

F = ALARM INDICATING APPLIANCE
T = TROUBLE SIGNAL

Figure 2-11. Typical remote station system.

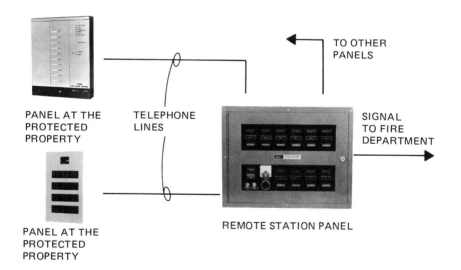

Figure 2-12. Basic components of a remote station protective signaling system.

have the capacity to operate the system for 60 hours, followed by 5 minutes of alarm. An independent trouble power supply must be provided for operation of trouble signals when the primary (main) power supply fails. The secondary (standby) supply may be used for this purpose.

Since a remote station system, like an auxiliary system, is intended primarily for property protection, an evacuation system is not required. If an evacuation system is installed, it should meet local requirements.

Remote station systems may have coded or noncoded operation. Coded systems can protect a maximum of five facilities, and may share common connections to the remote station. A noncoded system is used for alarm annunciation if only one facility is protected.

Proprietary Protective Signaling Systems

A proprietary protective signaling system is a signaling system located within a protected property, operated under single ownership, that provides emergency building control services. The proprietary system has a central supervising station staffed 24 hours a day. An alarm received in the central supervising station is transmitted to the fire department, and a permanent record of all alarms is maintained at the supervising station. (See Figure 2-13.)

The proprietary system is a widely used type of control unit in large commercial and industrial occupancies. Many existing proprietary systems have separate initiating device circuits for each building zone or subsection, similar to the local, auxiliary, and remote station systems. However, due to the increasing use of electronics, newer proprietary systems for large buildings often have signal multiplexing and built-in minicomputer systems. (See Figure 2-14.) These systems receive all signals throughout the building over one, or more, pair of wires, and determine the exact location of the fire by use of different frequencies or digitally coded information transmitted over the wires. Figure 2-15 shows the complexity and interrelationship of the many facets of a modern computer-controlled proprietary/high-rise communication system.

While a proprietary transmitting unit is similar to the other transmitting units, the receiving console can be quite different. A proprietary receiving console consists of individual lights, a digital display, or a cathode ray tube (CRT) visual display terminal and screen indicating

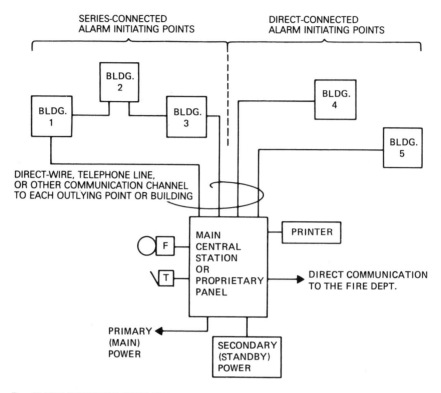

SERIES-CONNECTED
ALARM INITIATING POINTS

DIRECT-CONNECTED
ALARM INITIATING POINTS

BLDG. 2

BLDG. 1

BLDG. 3

BLDG. 4

BLDG. 5

DIRECT-WIRE, TELEPHONE LINE,
OR OTHER COMMUNICATION CHANNEL
TO EACH OUTLYING POINT OR BUILDING

F

MAIN
CENTRAL
STATION
OR
PROPRIETARY
PANEL

T

PRINTER

DIRECT COMMUNICATION
TO THE FIRE DEPT.

PRIMARY
(MAIN)
POWER

SECONDARY
(STANDBY)
POWER

F = ALARM INDICATING APPLIANCE
T = TROUBLE SIGNAL

Figure 2-13. Basic components of a proprietary protective signaling system.

the alarm point. An audible alarm to alert the console operator and a hard-copy printer can also be included.

Proprietary control units are required by NFPA 72, *National Fire Alarm Code*, to transmit trouble and alarm signals to the central supervising station. The control unit at the central supervising station, as well as remotely located control equipment, must have a secondary (standby) power supply that will operate the system for 24 hours under normal traffic conditions. Since operators are constantly on duty with a proprietary system, 24 hours of standby power is considered sufficient.

Large proprietary multiplex- and computer-controlled systems usually have functions in addition to alarm indication and initiation. These systems often provide for smoke control within the building by automatically closing and opening dampers in heating, ventilating, and air

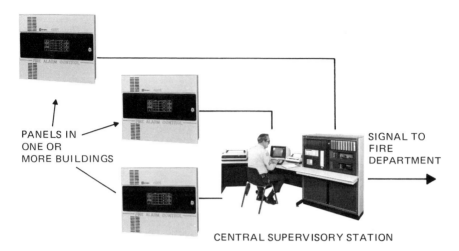

PANELS IN
ONE OR
MORE BUILDINGS

SIGNAL TO
FIRE
DEPARTMENT

CENTRAL SUPERVISORY STATION

PANELS IN CONTIGUOUS OR NONCONTIGUOUS PROPERTIES
UNDER ONE OWNERSHIP

Figure 2-14. Single-ownership properties can be protected by a proprietary system.

conditioning (HVAC) systems, and turning on exhaust fans. They can
also adjust elevator controls so elevators bypass fire floors and automatically return to the lobby floor for fire department use, as required by
ANSI A17.1, *Elevator Code.* In addition to increasing flexibility, use of
multiplexing signals greatly reduces the amount of building wiring.
Computer-based proprietary systems often include energy management
capabilities that result in energy savings, which has become a major factor in the increased use of these systems in large buildings.

The central supervising station for proprietary systems must be
staffed with competent personnel who are adequately trained to perform their duties. The supervising station must be constantly attended
by a minimum of one staff member, responsible for monitoring signals
and operating the system. "Dedicated personnel" is a main feature of a
proprietary system, and staff should not be assigned additional duties
that may interfere with these responsibilities.

Proprietary systems are used in facilities where a system more complex and elaborate than a local fire alarm system is needed. The central
supervising station must be located in a separate fire-resistive building
or in a fire-resistive enclosed room that is entirely separate from hazardous operations in the facility.

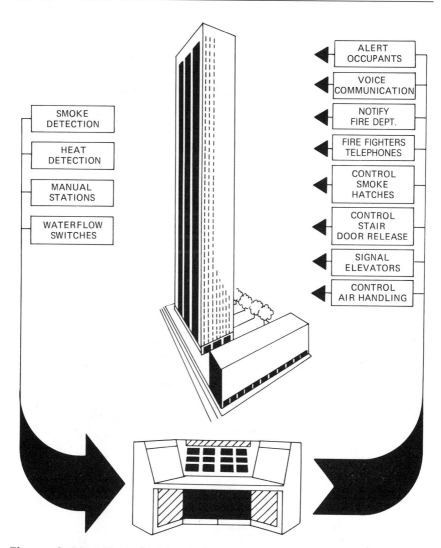

Figure 2-15. Typical high-rise fire alarm and communication system.

Fire alarms from the central supervising station to a fire department or fire brigade must be transmitted promptly and, preferably, automatically. Two-way telephone communication should also be established between the central supervising station and the fire department communication center.

Output signals from the central supervising station can control certain life safety and property protection functions in the building during emergencies. Control of building operations can include: elevator operation; unlocking stairwell and exit doors; releasing fire and smoke dampers; initiating and monitoring fire extinguishing systems and components; hazardous gas shutoffs; air pressurization of stairwells and means of egress; and control of lighting, HVAC systems, and processes.

Central Station Signaling System

NFPA 72, *National Fire Alarm Code*, defines a central station signaling system as a central alarm system connected to remote alarm and supervisory signaling devices attended by central station personnel 24 hours a day who monitor the system, investigate signals, and are responsible for system inspection, maintenance, and testing. (See Figure 2-16.)

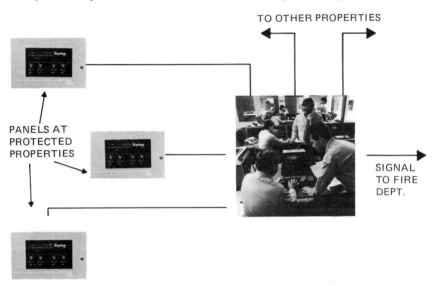

TO OTHER PROPERTIES

PANELS AT PROTECTED PROPERTIES

SIGNAL TO FIRE DEPT.

PANELS IN PROPERTIES NOT UNDER ONE OWNERSHIP

Figure 2-16. Multiowner properties connected in a central station system.

The central supervisory station that receives the fire alarm signal in a proprietary system is operated by an individual with a proprietary interest in the protected buildings. The central supervisory station of a proprietary system is generally located in a guard office within or near the building or group of buildings protected by the system. On the other hand, the supervising station in a central station signaling system is staffed by operators who perform the service for a fee and have no proprietary interest in the protected buildings.

In a central station signaling system, transmission usually occurs over two wires when a McCulloh circuit—the oldest transmitting means—is used. Transmission can be switched manually or automatically to transmit over one wire and ground. With this capability, the system will not be rendered inoperative by a break or ground fault in a single wire. Several facilities are customarily attached to a single transmission circuit. Each such circuit terminates in a recording instrument, and each subscriber has one or more coded signal numbers not repeated on that particular circuit.

A second means of signal transmission in central station systems is called the direct wire circuit. As its name implies, this is a direct dedicated (unswitched) pair of leased telephone lines running between the protected building and the central station panel. Limitations on circuit loadings and other means of transmission, such as requirements for digital alarm communicators and for cellular telephones, are specified in NFPA 72, *National Fire Alarm Code*.

When a signal is received at a central station, an alarm is immediately retransmitted to the public fire service communication center, and the appropriate authorities are informed. The central station dispatches someone to the protected building when equipment must be reset manually. Central station systems require secondary (standby) power supplies that will operate the system for a period of 24 hours.

Emergency Voice/Alarm Communication System

As defined by NFPA 72, *National Fire Alarm Code*, the purpose of an emergency voice/alarm communication system is to provide for the transmission of information and instructions pertaining to a fire emergency to the building occupants and fire department. This system also maintains communications with those persons remaining in the building. An emergency voice/alarm communication system can be used to partially evacuate a building in emergency conditions while permitting some building occupants to remain in safe areas. Such a system should always be provided whenever the emergency plan includes anything other than immediate and total evacuation of the building.

This system supplements a local, auxiliary, remote station, or proprietary protective signaling system. Its standby power supply must operate the voice/alarm signaling service for a fire or other emergency condition for 24 hours without the primary operating power and then be capable of operating the system in alarm condition for 5 minutes. Because of the emergency nature of a voice/alarm communication system, special requirements in NFPA 72, *National Fire Alarm Code*, also cover the survivability of the system so that fire damage to one paging zone will not result in loss of communication to another.

There are two basic types of emergency voice/alarm communication systems:

1. A stand-alone manually controlled system. Voice or tone signals for this type of system can be selected and distributed by the operator, and

2. A system that is integrated into a full fire alarm system. Voice or tone signals can be selected and distributed manually or automatically for the integrated system.

The voice/alarm system consists of a series of high-reliability speakers located throughout the building. (See Figure 2-17.) They are connected to and controlled from the fire alarm communication console located in an area of the building designated as a fire command headquarters. From the fire command headquarters, voice messages can be provided to individual speaker zones or the entire building with specific instructions to the occupants. Some systems have fire warden stations, or fire zones, on each floor from which a fire warden assumes local command and passes on specific evacuation instructions. Operation of the command headquarters usually is assumed by a trained building employee until the fire department arrives, at which time the officer in charge takes over. The system also can be used during fire fighting operations for communication with the fire fighters.

A one-way system permits emergency personnel to give voice instructions on either a selective or on an "all call" basis via microphone and a loudspeaker system. Its use began with the advent of high-rise buildings and other structures where immediate evacuation of the entire building might be impractical, and the fire plan called for selective partial evacuation or the relocation of the building occupants from the fire area.

A two-way system, using telephone or intercom techniques, permits communication between fire service personnel while they are investigating or fighting a fire in the building. Most two-way systems use telephones located in fire stairwells so fire service personnel can be assured of reliable communication from any floor to the fire command station.

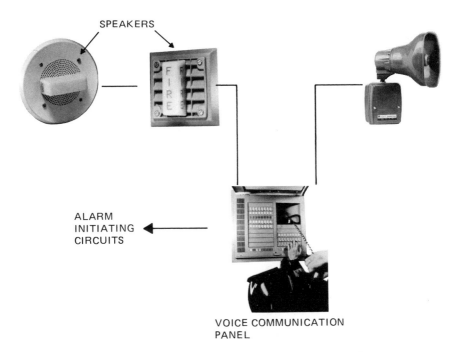

Figure 2-17. Typical speakers connected to voice panel in emergency voice/alarm system.

These telephones are also permitted to be used by other building occupants and are frequently installed on all floors for use in emergency circumstances. A distinctive notification signal, different from any other alarm or trouble signal, is required at the fire command station to indicate an off-hook condition of any telephone on the system.

One important aspect of a voice/alarm communication system is that, since complete building evacuation is not always feasible, occupants can be instructed to relocate to "safe" areas and wait out the fire. (See Chapter 7, "Signal Notification.")

CONCLUSION

The type of protective signaling system to be used depends on the protection needs, i.e., life safety and/or property protection, of each facility or individual property. While there are requirements basic to all protective signaling (fire alarm) systems, specific features of each system—type of primary and secondary power supply, trouble signals, method of signal transmission, system supervision, etc.—must be carefully considered by the designer for each fire alarm system. NFPA 72, *National Fire Alarm Code*, should be consulted when the designer is "matching" the protective system to the protection use desired.

BIBLIOGRAPHY

NFPA Codes, Standards, Recommended Practices, and Manuals. (See the latest *NFPA Catalog* for availability of current editions of the following documents.)

NFPA 70, *National Electrical Code*.

NFPA 72, *National Fire Alarm Code*.

NFPA 1221, *Standard for the Installation, Maintenance, and Use of Public Fire Service Communication Systems*.

Additional Reading

ANSI A17.1, *Elevator Code*, American National Standards Institute, 11 W. 42nd St.—13th Floor, New York, NY 10036.

3

Signal Initiation

INTRODUCTION

Fire alarm signals can be initiated by several types of devices (either manual or automatic) or through a "systems" approach in which alarm signals are combined with functions of fire extinguishing systems. Systems or individual devices are activated (and a signal initiated) by heat, smoke, energy radiation, or other detectable byproducts of a fire.

An automatic initiating device must be able to differentiate between normal environmental fluctuations, i.e., nonfire conditions, and prefire conditions or changes. A signal is initiated when certain preset conditions (usually related to temperature, gas, light, or energy) are reached or exceeded.

Initiating devices can be used in a facility in combination with a fire extinguishing system. Extinguishing system capacity is determined by the area to be protected and the speed of detection desired. Extinguishing systems that are connected to alarm systems require supervisory devices to detect abnormal water pressure, water tank temperature, and water level changes. Signals can also be initiated via a guard's tour supervisory service.

For additional information on audible signals see Chapter 7, "Signal Notification"; on design parameters for heat and smoke detectors, see Chapter 9, "Fire Detection System Design"; also refer to Chapter 13, "Installation."

ALARM INITIATING DEVICES

Alarm initiating devices for fire alarm systems are either manual fire alarm boxes (stations) or automatic fire detectors. Both are used to initiate an alarm on a fire alarm system.

General Requirements for Manual Stations: Mounting and Distribution

Manual fire alarm boxes should be approved for the particular application and can be used for fire protective signaling purposes only. However, combination fire alarm and guard's tour boxes are acceptable. Each box should be securely mounted with the bottom of the box not less than 3.5 ft (1.1 m) or more than 5 ft (1.4 m) above the floor level.

Manual fire alarm boxes should be distributed throughout the protected area so they are unobstructed, readily accessible, and located in the normal path of exit from the area, as follows:

1. At least one box provided on the first floor and on each floor.
2. Additional boxes should be provided so that travel distance to the nearest box will not exceed 200 ft (61 m).

Manual Fire Alarm Boxes

Manual stations can be one of four types: (1) noncoded or coded, (2) presignal or general alarm, (3) breakglass or nonbreakglass, and (4) single or double action.

Noncoded or Coded Stations: A noncoded manual station contains a normally open or closed switch that is housed within a distinctive enclosure. Once actuated, the station contacts transfer. This condition is maintained until the station is reset to normal. Older coded stations contain a mechanically or electrically driven motor that, when activated, turns a code wheel, causing contacts to momentarily open or close to reproduce the code of the station. Newer designs are electronic but function identically. The station is required to repeat its code a minimum of three times. Both coded and noncoded stations can be surface, semiflush, or flush mounted.

Presignal or General Alarm Stations: Presignal stations initially cause alarm signals to sound only in specific areas. Actuation of a key switch on the station or the control panel will cause an evacuation signal to sound. A general alarm station is the most common manual alarm system. When actuated, a general alarm station causes evacuation signals to sound immediately.

Breakglass or Nonbreakglass Stations: The term "breakglass" is applied to both noncoded and coded stations. For a breakglass station,

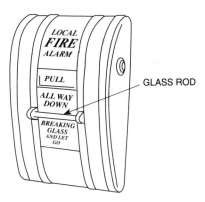

Figure 3-1. Breakglass station.

actuation of the device to cause an alarm requires the initial action of breaking a glass or other breakable element. This is thought to deter false alarms, and is also intended to address the requirement that, once activated, manual stations cannot be reset without use of a special tool. Stations without this feature are classified as "nonbreakglass." (See Figure 3-1.)

Single- or Double-Action Stations: A single-action station requires only one action by the user to initiate an alarm. The required action is usually breaking a glass element or actuating a lever or other moveable part of the station. (See Figure 3-2.)

A double-action station initiates an alarm as a result of two actions taken by the user. Typically the user must break a glass, open a door, or lift a cover to gain access to a switch or lever that must then be operated as the second action to initiate an alarm. (See Figure 3-3.)

Automatic Fire Detectors

People are excellent but unreliable fire detectors. Various mechanical, electrical, and electronic devices in use today mimic the human senses of smell, sight, hearing, taste, and touch to detect the environmental changes created by fire.

The most common elements of a fire that can be detected are heat, smoke (aerosol particulate), gas, and light radiation. However, not all fires produce all of these elements, and nonfire conditions can produce similar ambient conditions. The fire alarm signaling system designer must be able to differentiate among those elements that might be

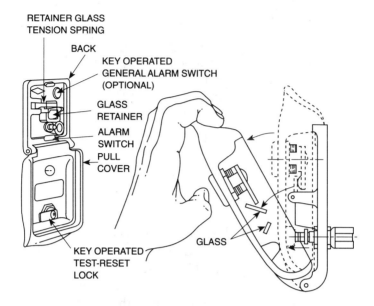

Figure 3-2. Single-action station.

Figure 3-3. Double-action station.

expected from a fire. The designer must also know the similar ambient conditions that might result from nonfire situations.

Even if all of the elements of fire--heat, smoke, and light--are present in a given fire, the magnitude of each element must exceed some theoretical basic level during fire development. It should be noted that one element will usually appear first; this is especially important if the system is designed primarily for life safety.

Automatic fire detectors are classified in Chapter 5 of NFPA 72, *National Fire Alarm Code*, as heat-, smoke-, radiant energy-, and fire-gas-sensing detectors, and "other" fire detectors. A heat detector is a device that detects abnormally high temperature or rate-of-temperature rise. Smoke detectors are devices that detect the particles of combustion. A flame detector detects the infrared, ultraviolet, or visible radiation produced by a fire. Fire-gas detectors are devices that detect gases produced by a fire. "Other" fire detectors are devices that detect phenomena other than heat, smoke, flame, or gases produced by a fire.

Detectors are further divided into three types: (1) line type, (2) spot type, and (3) air-sampling type. In a line-type detector, detection is continuous along a path. Typical examples of line-type devices are rate-of-rise pneumatic tubing detectors, projected beam smoke detectors, and heat-sensitive cable detectors.

The detecting element is concentrated at a particular location in a spot-type detector. Typical examples of spot-type detectors are bimetallic detectors, fusible alloy detectors, certain pneumatic rate-of-rise detectors, most smoke detectors, and thermoelectric detectors.

An air-sampling-type detector consists of piping or tubing distribution from the detector unit to the area(s) to be protected. An air pump draws air from the protected area back to the detector through the air-sampling ports and piping or tubing. The air is analyzed for fire products at the detector, with current examples employing either photoelectric (scattering) or cloud chamber principles.

Automatic detectors have two operating modes: (1) nonrestorable and (2) restorable. A nonrestorable detector is a device whose sensing element is designed to be destroyed in the process of detecting a fire. In a restorable detector the sensing element is not ordinarily destroyed in the process of detecting a fire. Restoration of the element may be manual or automatic.

In addition, detectors that respond in more than one way to a fire condition are called combination detectors. Combination detectors either: (1) respond to more than one of the fire phenomena (heat, smoke, flame, etc.) or (2) employ more than one operating principle to sense one of these phenomena. Two typical examples of combination detectors are: (1) a combination of a heat detector with a smoke detector or (2) a combination rate-of-rise and fixed temperature heat detector.

Heat Detectors: Heat detectors are the oldest type of automatic fire detection device, ranging from the development of automatic sprinklers in the 1860s and continuing to the present with various types of devices. A sprinkler can be considered a combined heat-activated fire detector and extinguishing device when the sprinkler system is provided with waterflow indicators connected to the fire alarm control system. Waterflow indicators detect either the flow of water through the pipes or the subsequent water pressure change upon actuation of the system.

Heat detectors that only initiate an alarm and have no extinguishing function are still in limited use. Although they are the least expensive of fire detectors and have the lowest false alarm rate of all automatic detecting devices, they also are the slowest in detecting fires. A heat detector is best suited for fire detection in a small confined space where rapidly building high heat output fires are expected, in areas where ambient conditions would not allow the use of other fire detection devices, or where speed of detection is not a prime consideration.

Heat detectors are generally located on or near the ceiling and respond to the convected thermal energy of a fire. They respond either to a specified rate of temperature change or when the detecting element reaches a predetermined fixed temperature. In general, heat detectors are designed to operate when heat causes a prescribed change in a physical or electrical property of a material or gas.

• *Operating Principles of Fixed Temperature Heat Detectors:* Fixed temperature heat detectors are designed to initiate an alarm when the temperature of the operating element reaches a specific point. The air temperature at the time of alarm is usually considerably higher than the rated temperature, because it takes time for the air to raise the temperature of the operating element to its set point. This condition is called thermal lag. Fixed temperature heat detectors are available to cover a wide range of operating temperatures from approximately 100°F (38°C) and higher. (See Table 3-1.) Higher temperature detectors are necessary so that detection can be provided in areas normally subjected to high ambient (nonfire) temperatures, or in areas zoned so that only detectors in the immediate fire area operate.

TABLE 3-1. Operating Temperature Ranges for Fixed Temperature Heat Detectors

Temperature Classification	Temp. Rating Range (°F)	Max. Ceiling Temp. (°F)	Color Code
Low*	100 to 134	20 below†	Uncolored
Ordinary	135 to 174	100	Uncolored
Intermediate	175 to 249	150	White
High	250 to 324	225	Blue
Extra High	325 to 399	300	Red
Very Extra High	400 to 499	375	Green
Ultra High	500 to 575	475	Orange

*Intended only for installation in controlled ambient areas. Units marked to indicate maximum ambient installation temperature.

†Maximum ceiling temperature has to be 20°F or more below detector-rated temperature.

NOTE: The difference between the rated temperature and the maximum ambient temperature should be as small as possible to minimize the response time.

For SI Units: $°C = 5/9 \ (°F{-}32)$.

• *Nonrestorable Fusible Element-Type Detectors:* Eutectic metals (alloys of bismuth, lead, tin, and cadmium that melt rapidly at a predetermined temperature) can be used as operating elements for heat detection. The most common such use for these metals is the fusible soldered joint element in an automatic sprinkler. (See Figure 3-4.) These elements are nonrestorable, since fusing the element allows the valve cap covering the sprinkler orifice to fall away, water to flow in the system, and the alarm to be initiated.

A closed-head sprinkler can be used as a release device on a fixed temperature detector for pneumatic operation on a hydraulic differential deluge valve. The closed-head sprinklers should be attached to ½-in. (12.7-mm) diameter air pilot lines on approximately 10-ft (3-m) spacing. The dry pilot actuation release system is pressurized with compressed air or nitrogen. Upon activation of a sprinkler, release of pressure from the pilot lines activates the deluge valve. To increase the speed of operation of the deluge valve, an accelerator can be installed on the pilot line detection system. Typical applications of the pneumatic pilot line detection system are transformer protection and cooling tower protection. The pneumatic detection system can also be utilized

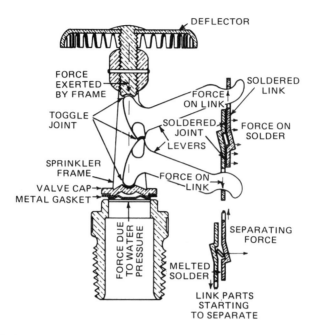

Figure 3-4. Representative arrangement of a soldered link and lever automatic sprinkler.

in those areas where hazardous electrical operations are performed requiring deluge system protection.

A eutectic metal may also actuate an electrical heat detector and is often used as a solder to secure a spring under tension. When the element fuses, the spring action closes contacts and initiates an alarm. (See Figure 3-5, items D, F, and G.) Devices using eutectic metals cannot be restored; either the device or its operating element must be replaced following operation.

• *Restorable Bimetallic-Type Detectors:* In a restorable bimetallic-type detector, two metals with different coefficients of thermal expansion are bonded together and then heated. Differential expansion causes bending or flexing toward the metal with the lower expansion rate. This action closes a normally open circuit and initiates an alarm. The low-expansion metal commonly used is Invar™, an alloy of 36 percent nickel and 64 percent iron. Several alloys of manganese-copper-nickel, nickel-chromium-iron, or stainless steel may also be used for the high-expansion component of a bimetal assembly. Bimetals are used for the

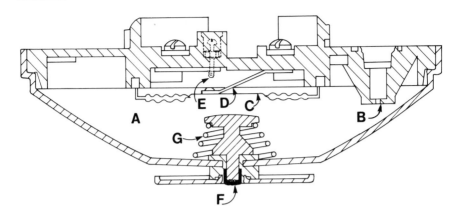

Figure 3-5. A spot-type combination rate-of-rise, fixed temperature device. The air in chamber A expands more rapidly than it can escape from vent B. This causes pressure to close electrical contact D between diaphragm C and contact screw E. Fixed temperature operation occurs when fusible alloy F melts releasing spring G which depresses the diaphragm closing contact points.

operating elements as either strips or snap discs for fixed temperature detectors.

Figure 3-6 shows a bimetal strip detector, where Metal A changes its length when heated. Conversely, a different metal, Metal B, has a higher coefficient of expansion and will expand more than Metal A when the temperature rises. When these two metals are bonded together and the temperature changes from a lower to a higher temperature, Metal B (which is on the bottom) will expand more than Metal A, causing the entire element to bend upward as shown in the lower part of the figure. This bimetallic action closes an electrical circuit as shown in Figure 3-7 and is similar in operation to that of a bimetallic thermostat controlling the furnace in a home.

A bimetal strip, as it is heated, deforms in the direction of the contact point. With a given bimetal, the width of the gap between the contacts determines the operating temperature; the wider the gap, the higher the operating point.

The disc-type bimetallic element (Figure 3-7, bottom) is also used extensively in fire alarm systems. The round disc is formed of bimetallic material and is prestressed in an arch shape. When the temperature of the bimetallic material rises to a predetermined point, the expansion of the metals causes the disc to change from a concave to a convex

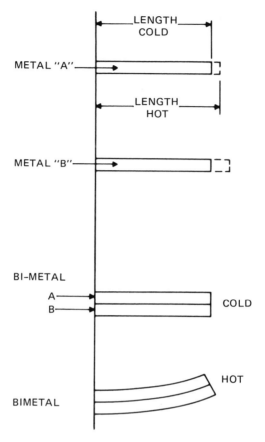

Figure 3-6. Bending action with two dissimilar metals.

position in a snap action (as shown by the dotted line). Because of the snap action, the force applied to the contact points is greater on initial contact than the force applied by strip-type bimetallic elements.

Bimetallic detectors are usually designed to restore themselves to their original condition (and the disc to its concave position) after the heat has been removed. (See Figure 3-8 and Figure 3-9).

A heat collector usually is attached to the frame of the detector to transfer heat more quickly from the room air to the bimetal.

Heat detectors with bimetallic elements are automatically self-restoring after operation when the temperature drops below the operating (set) point. With a bimetallic strip element, the detector will usually be designed in such a manner that the detector restores when the element is lowered just a degree or so below the set point. However, if

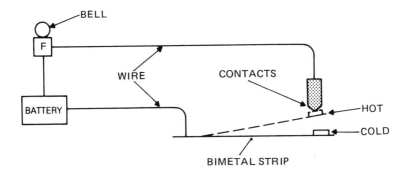

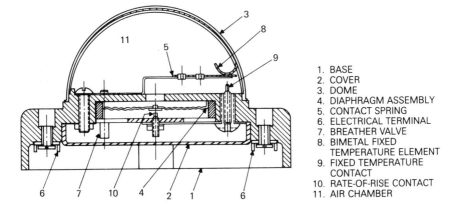

Figure 3-7. (Top) Bending action in bimetallic strip detector; (bottom) bimetallic disc-type detector.

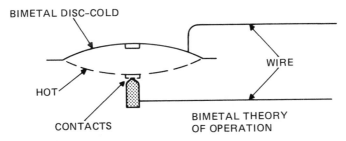

1. BASE
2. COVER
3. DOME
4. DIAPHRAGM ASSEMBLY
5. CONTACT SPRING
6. ELECTRICAL TERMINAL
7. BREATHER VALVE
8. BIMETAL FIXED
   TEMPERATURE ELEMENT
9. FIXED TEMPERATURE
   CONTACT
10. RATE-OF-RISE CONTACT
11. AIR CHAMBER

Figure 3-8. Combination rate-of-rise and fixed temperature detector (with resetting fixed temperature element).

the element is a snap disc, the ambient temperature around the detector may need to be lowered well below the set point for the disc element to snap back to its concave position. For this reason, and the 15°F (8.3°C)

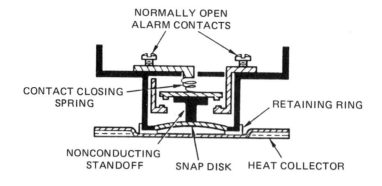

Figure 3-9. Spot-type fixed temperature snap-disc detector.

operating tolerance permitted for a typical 135°F (57.5°C) ordinary-degree fixed temperature heat detector, a detector of this temperature rating should not be installed where the ceiling temperature might exceed 100°F (37.8°C). (See Table 3-1 for other detector temperature ratings.)

• *Line-Type Heat Detectors:* Various methods of line-type detection have been developed as alternatives to spot-type fixed temperature detection, such as the heat-sensitive cable continuous line-type detector. The detector shown in Figure 3-10 uses a pair of steel wires in a normally open circuit. The conductors are held apart by heat-sensitive insulation. The wires, under tension, are enclosed in a braided sheath to form a single cable assembly. When the temperature limit is reached, the insulation melts, the two wires contact, and an alarm is initiated.

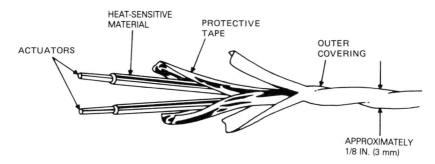

Figure 3-10. Line-type heat detector.

The fused section of the cable must be replaced following an alarm to restore the system.

A similar alarm device utilizing a semiconductor material and a stainless steel capillary tube has been used where mechanical stability is a factor in heat detection. (See Figure 3-11.) The capillary tube contains a coaxial center conductor separated from the tube wall by a temperature-sensitive semiconductor material. Under normal conditions, a small current (below the alarm threshold) flows in the circuit. As the temperature rises, the resistance of the semiconductor thermistor decreases, allows more current flow, and initiates the alarm.

• *Rate Compensation-Type Detectors:* A rate compensation detector is a device that responds when the temperature of the surrounding air reaches a predetermined level, regardless of the rate-of-temperature rise. (See Figure 3-12.)

A typical example of a rate compensation detector is a spot-type detector. This type of detector has a tubular casing of a metal that tends to expand lengthwise as it is heated, and an associated contact mechanism that will close at a certain point in the elongation. A second metallic element inside the tube exerts an opposing force on the contacts, tending to hold them open. The forces are balanced in such a way that on a slow rate-of-temperature rise, heat takes longer to penetrate to the inner element. This inhibits contact closure until the total device has been heated to its rated temperature level. However, on a fast rate-of-temperature rise, there is not as much time for heat to penetrate to the inner element, which therefore exerts less of an inhibiting effect. This contact closure is obtained when the total device has been heated to a lower level, compensating for thermal lag.

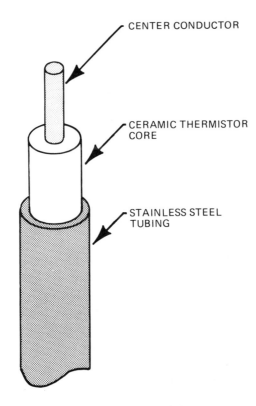

Figure 3-11. View of the construction of the continuous thermal sensor showing outer tubing, ceramic thermistor core, and center wire.

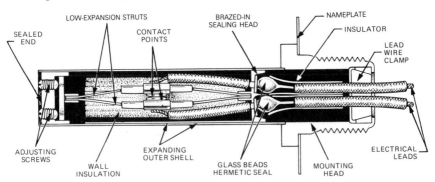

Figure 3-12. Section of spot-type rate compensation detector.

Rate compensating detectors are also automatically self-restoring after operation when the ambient temperature drops to some point below the operating point.

• *Rate-of-Rise-Type Detectors:* One effect that a flaming fire has on the surrounding area is to rapidly increase air temperature in the space above the fire. Fixed temperature heat detectors will not initiate an alarm until the air temperature near the ceiling exceeds the design operating point. The rate-of-rise detector, however, will function when the rate of temperature increase exceeds a predetermined value, typically around 12 to 15°F (7 to 8°C) per minute. Rate-of-rise detectors are designed to compensate for the normal changes in ambient temperature [less than 12°F (7°C) per minute] that are expected under nonfire conditions.

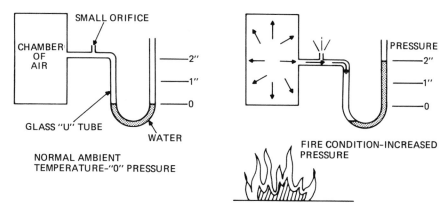

Figure 3-13. Operating principle for pneumatic fire detector.

In a pneumatic fire detector, air heated in a tube or chamber expands, increasing the pressure in the tube or chamber. (See Figure 3-13.) This exerts a mechanical force on a diaphragm that closes the alarm contacts. If the tube or chamber were hermetically sealed, slow increases in ambient temperature, a drop in the barometric pressure, or both would cause the detector to initiate an alarm regardless of the rate of temperature change. To overcome this, pneumatic detectors have a small orifice to vent the higher pressure that builds up during slow increases in temperature or a drop in barometric pressure. The vents are sized in such a manner that when the temperature changes rapidly, as in a fire situation, the rate of expansion exceeds the venting rate and the pressure rises. When the temperature rise exceeds 12 to 15°F (7 to 8°C) per minute, a flexible diaphragm converts the pressure to mechanical action. Pneumatic heat detectors are available for both line- and spot-type detectors. Rate-of-rise operation for a line-type pneumatic heat detector is shown in Figure 3-14, and a spot type in Figures 3-5, 3-9, and 3-15.

Most spot-type rate-of-rise detectors are equipped with a fixed temperature backup feature so that should the temperature rise be slower than the 15°F (8°C) per minute rate, the detector will operate when the detecting element has reached the predetermined fixed point. Spot-type rate-of-rise detectors are installed in the same manner as fixed temperature detectors.

The line-type pneumatic heat detector consists of metal tubing in a loop configuration attached to the ceiling or side wall near the ceiling of the area to be protected. Lines of tubing are normally spaced not more than 30 ft (9.1 m) apart, not more than 15 ft (4.5 m) from a wall, and with no more than 1,000 ft (305 m) of tubing on each circuit. Also, a minimum of at least 5 percent of each tube circuit or 25 ft (7.6 m) of tube, whichever is greater, must be in each protected area. Without this minimum amount of tubing exposed to a fire condition, insufficient pressure would build up to achieve proper response.

Shock chambers tend to compensate for additional short but unsustained increases in pressure that might be caused, for example, by the opening of an oven door. This reduces the possibility of a false alarm.

In small areas, where line-type detectors might have insufficient tubing exposed to generate sufficient pressures to close the alarm contacts, air chambers or rosettes of tubing are often used. These units act like a spot-type detector by providing the volume of air required to meet the 5 percent or 25-ft (7.6-m) requirement. Since a line-type rate-of-rise detector is an integrating detector, it will actuate either when a rapid heat rise occurs in one area of exposed tubing, or when a slightly less rapid heat rise takes place in several areas where tubing on the same loop is exposed.

The pneumatic principle is also used to close contacts within spot-type detectors. The difference between line- and spot-type detectors is that the spot type contains all of the air in a single container rather than in a tube that extends from the detector assembly to the protected area(s).

• Sealed Pneumatic Line-Type Detectors: The sealed pneumatic line-type heat detector is another fixed temperature heat detector for special applications. It is not vented to the air and consists of a capillary tube containing a special salt that is saturated with hydrogen gas. At normal temperatures, most of the hydrogen is held in the porous salt, and the pressure in the tube is low.

As the temperature increases at any point along the tubing, hydrogen is released from the salt which increases the internal pressure and even-

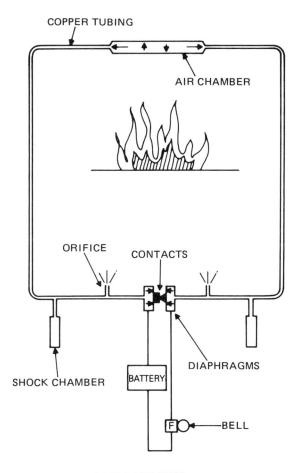

F = ALARM INDICATING APPLIANCE

Figure 3-14. Operating principle for line-type pneumatic heat detector.

tually trips a diaphragm pressure switch. The integrity of the capillary tube is supervised by a second pressure switch that monitors the low pressure present at normal temperatures. (See Figure 3-16.)

• *Thermoelectric Effect Detectors:* A thermoelectric effect detector is a device with a sensing element consisting of a thermocouple or thermopile unit that produces an increase in electric voltage in response to an increase in temperature. This potential is monitored by associated control equipment, and an alarm is initiated when the voltage increases at an abnormal rate. (See Figure 3-17.)

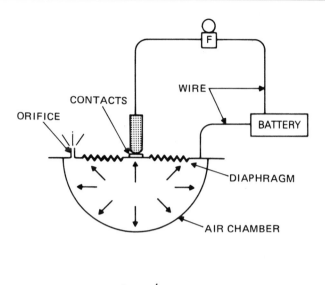

F = ALARM INDICATING APPLIANCE

Figure 3-15.   Operating principle for spot-type pneumatic heat detector.

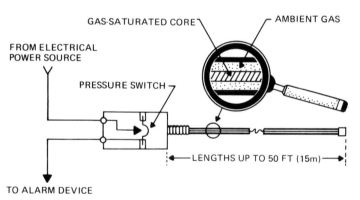

Figure 3-16.   Sealed line-type fixed temperature heat detector.

Thermopile devices, which operate in the voltage-generating mode, use two sets of thermocouples. One set is exposed to changes in the atmospheric temperature. When temperature rapidly changes as in a fire situation, the temperature of the exposed set increases faster than

Figure 3-17. Spot-type thermoelectric effect heat detector.

the temperature of the unexposed set, generating a net voltage. The voltage increase associated with this potential operates the alarm circuit. Since the thermopile units are connected in series, this voltage need not be within only one detector; the small voltages produced at each unit on a circuit can cumulatively produce an alarm. The sensitivity of a thermopile detector is directly related to the number of thermocouple junctions within the device; the sensitivity increases as the number of junctions increases. Sensitivity can also be increased by designs that focus radiative energy on the exposed junctions.

● *Combination Detectors:* Combination detectors contain more than one fire-responsive element. These detectors may be designed to respond from either element, or from the combined partial/complete response of both elements. An example of an either/or response element is a heat detector that operates on both the rate-of-rise and fixed temperature principles. The advantage of this type of detector is that the rate-of-rise element will respond quickly to a rapidly developing fire, while the fixed temperature element responds to a slowly developing fire when the detecting element reaches its set point temperature. The most common combination detector uses a vented air chamber and a flexible diaphragm for the rate-of-rise function, while the fixed temperature element is usually a leaf-spring restrained by a eutectic metal. When the fixed temperature element reaches its operating temperature, the eutectic metal fuses and releases the spring, closing the contacts.

Smoke Detectors: A smoke detector will detect most fires much more rapidly than a heat detector. Smoke detectors are identified by their operating principle; two of the most common operating principles are: (1) ionization and (2) photoelectric. As a class, smoke detectors using the ionization principle provide somewhat faster response to high-energy (open flaming) fires, since these fires produce large numbers of smaller smoke particles. As a class, smoke detectors operating on the photoelectric principle respond faster to the smoke generated by low-energy (smoldering) fires, as these fires generally produce more of the larger smoke particles. However, each type of smoke detector is subjected to, and must pass, the same test fires for listing at testing laboratories.

Smoke detectors can also initiate control of smoke spread in four different ways: (1) by prevention of the recirculation of dangerous quantities of smoke within a building, (2) selective operation of equipment to exhaust smoke from a building, (3) selective operation of equipment to pressurize smoke compartments, and (4) operation of doors to close the openings in smoke compartments.

Cloud chamber, air duct, and gas-sensing detectors are also based on smoke particle principles, and will be discussed herein.

• *Ionization-Type Smoke Detectors:* An ionization smoke detector (usually the spot type) has a small amount of radioactive material in the sensing chamber. This material ionizes the air, rendering the air conductive and permitting a current flow through the air between two charged electrodes. This gives the sensing chamber an effective electrical conductance. When smoke particles enter the ionization area, they decrease the conductance of the air by attaching themselves to the ions, causing a reduction in ion mobility. The detector responds when the conductance is below a predetermined level.

The theory of ionization detector operation is based on alpha particle properties. Alpha particles ionize air, dissociating the air molecules into positive ions and negative electrons. If this ionized air is introduced into an electric field, the charges on the ions and the electrons will cause them to move and generate a current. As shown in Figure 3-18, a potential from Battery B is applied to Plates P_1 and P_2. The air between the plates is ionized by the Alpha Emitter A, and the charged particles move in the direction indicated by the arrows. A galvanometer measures the current; the value of the current depends upon the strength of the alpha emitter and to a lesser extent on the voltage of Battery B. The current in an ionization chamber depends upon the composition of the gas between the electrodes. Ion production is dependent on the number

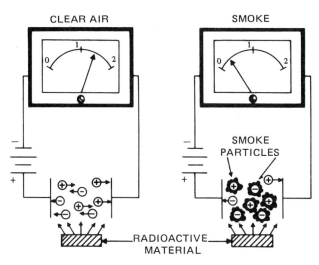

Figure 3-18. Principle of operation for ionization smoke detector.

and size of the gas molecules. The rate of drift of the ions is closely related to their size and mass, and is not based on gas composition.

Conditions are quite different when the smoke particles enter the ionization chamber. These particles attach themselves to the ions, causing a sharp drop in current that effectively increases the resistance of the chamber. This, in turn, sounds the alarm through other electrical circuits.

A cross section view of an ionization detector is shown in Figure 3-19.

• *Photoelectric-Type Smoke Detectors:* The suspended smoke particles generated during the combustion process affect the propagation of a light beam passing through the air. This effect can be utilized to detect the presence of a fire in two ways: (1) obscuration of light intensity over the beam path or (2) scattering of the light beam.

Smoke detectors that operate on the light obscuration principle consist of a light source, a light beam collimating system, and a photosensitive device. When dense smoke obscures part of the light beam, or less-dense smoke obscures more of the beam, the amount of light reaching the photosensitive device is reduced by a combination of absorption and scattering, initiating the alarm. (See Figure 3-20.) The light source is usually a light-emitting diode (LED), which is a reliable long-life source of illumination with a low current requirement. Pulsed LEDs can generate sufficient light intensity while operating at low overall power levels for use in detection equipment.

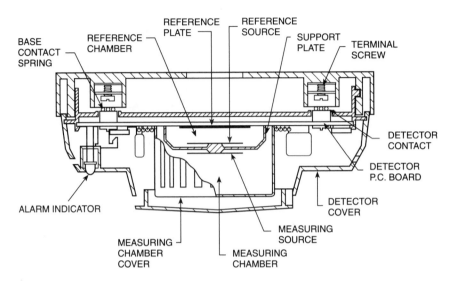

Figure 3-19.   Cross section view of an ionization smoke detector.

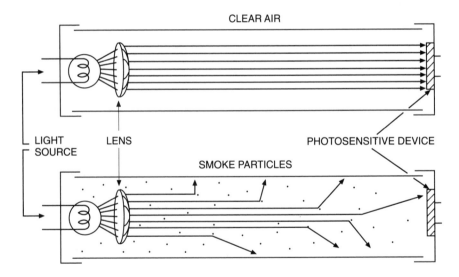

Figure 3-20.   Principle of operation for photoelectric obscuration smoke detector.

Most light obscuration smoke detectors are the beam type and are used to protect large open areas. They are installed with the light source at one end of the area to be protected, and the photosensitive device at the other. In some applications, mirrors determine the area of coverage by directing the beam over the desired path. For each mirror used, however, the rated beam length of the device must be progressively reduced by one-third, as illustrated in Table 3-2. Projected beam detectors are generally installed within 20 in. (51 cm) of, and parallel to, the ceiling.

TABLE 3-2. Projected Beam Using Mirrors

Number of Mirrors		Maximum Allowable Beam Length
0	Listed Length L	
1	2/3 L = a + b	
2	4/9 L = c + d + e	

EXAMPLE: Maximum allowable length of beam listed for 300 ft (L) using two mirrors is 4/9 × 300 ft, or 133 ft.

For SI Units: 1 ft = 0.30 m.

Light scattering results when smoke particles enter a light path. Smoke detectors utilizing this photoelectric principle are usually of the spot type and contain a light source and a photosensitive device arranged so light rays normally do not fall onto the photosensitive device. When smoke particles enter the light path, light strikes the particles and is scattered onto the photosensitive device, causing the detector to respond. (See Figure 3-21.) A photodiode or phototransistor is usually the photosensitive device used in light scattering detectors.

• *Cloud Chamber Smoke Detectors:* In a cloud chamber-type smoke detector, the concentration of submicrometer particles is measured on a "one particle-one droplet" basis by photoelectric (attenuation) methods. An air sample containing submicrometer smoke particles is drawn through a humidifier where it is brought to 100 percent relative humidity. The sample then passes to an expansion chamber where the pressure is reduced with a vacuum pump, causing condensation of water on the particles. The droplets quickly grow to a visible size. A light beam

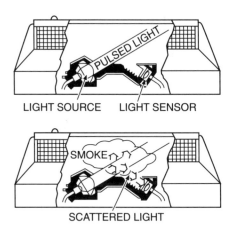

Figure 3-21. Principle of operation for photoelectric scattering smoke detector; cross section view of detector. (Top: clean-air; bottom, with smoke.)

passing through the cloud of water droplets is used to measure the concentration of droplets formed. The output voltage from the light receiver is directly proportional to the number of droplets (i.e., the number of condensation nuclei present).

The cloud chamber system uses a mechanical valve and switching arrangement to allow sampling from up to four detection zones, with as many as ten sampling heads per zone. (See Figure 3-22.) Each zone is sampled once per second for 15 seconds; all four zones are sampled each minute.

• *Air-Sampling Smoke Detectors:* Optical air monitoring equipment has been configured to provide another technique for early warning fire detection. These particle detectors sample the air from a protected area. These detectors are capable of protecting large areas because of the inherent sensitivity of the devices. In addition, the detectors can be used in areas having high air change rates where dilute smoke concentrations or laminar airflows interfere with the proper operation of other types of smoke detectors. These air-sampling detectors can draw air through a piping network to the detector unit by an air-aspirating fan in the detector assembly. Air samples are illuminated with a high-intensity strobe light, which causes smoke particles to reflect light to a solid-state photo receiver. An analog signal is generated from the detector to the control unit that displays the smoke obscuration sensitivity. The detector system provides independent programmable levels of alarms to indicate different levels of fire conditions. The two main advantages of this

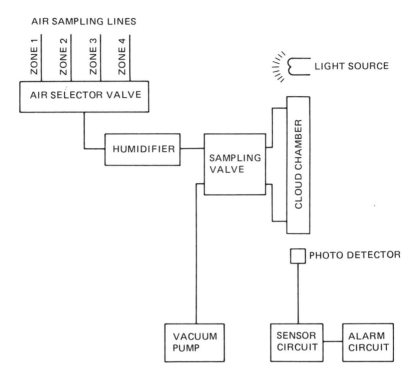

Figure 3-22. Cloud chamber smoke detector.

detector are: (1) the sensitivity settings for incipient fire detection, and (2) one detector apparatus can cover a relatively large area, ranging up to 20,000 sq ft (1800 m^2), by using perforated piping for air sampling in the protected area.

• *Air Duct Smoke Detectors:* Air duct smoke detectors function by detecting smoke to control air movement by air conditioning and ventilating systems. Detectors installed in an air duct system should not be used as a substitute for open-area protection, because smoke may not be drawn from open areas when air-conditioning systems or ventilating systems are shut down. Further, smoke-laden air can be diluted by clean air from other parts of the building or by outside air intakes, which may allow high densities of smoke in a single room with no appreciable smoke in the air duct at the detector location.

Air duct smoke detection has definite limitations. It is not a substitute for an area smoke detector or early warning detection, and it is not a replacement for a building's regular fire detection system. Duct smoke detectors normally sample great volumes of air from large areas of coverage and cannot be expected to match the detection ability of area detectors. It should also be noted that dirt-contaminated air filters can restrict air flow, causing a reduction in the operating effectiveness of the duct smoke detectors. Area smoke detectors are the preferred means of controlling smoke spread, because duct smoke detectors can only detect smoke when smoke-laden air is circulating in the ductwork. Fans may not be running at all times, e.g., during cyclical operation or during temporary power failure.

Air duct detectors should be listed for the purpose for which they are used, and securely installed in such a way as to obtain a representative sample of the air stream. Secure installation can be accomplished by:

1. Rigid mounting within the duct (see Figure 3-23),
2. Rigid mounting to the interior wall of the duct, with the sensing element protruding into the duct,
3. Placement outside the duct, with rigidly mounted sampling tubes protruding into the duct (see Figure 3-24), or
4. Use of a projected light beam through the duct.

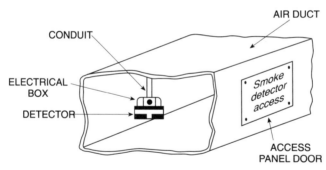

Figure 3-23.   Air duct detector mounted within the duct.

Air duct smoke detector design is based on one of the following operation principles: ionization, photoelectric, or cloud chamber.

Gas-Sensing Fire Detectors: Many changes occur in the gas content of the environment during a fire, especially the addition of gases that are not usually present in such concentrations. For example, higher than usual levels of gases, such as H_2O, CO, CO_2, HCl, HCN, HF, H_2S, NH_3, and various oxides of nitrogen, can be present in a fire. With the exception of water (H_2O), carbon monoxide (CO), and carbon dioxide (CO_2),

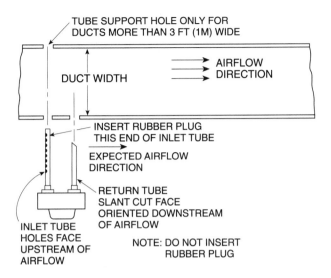

Figure 3-24. Procedure for rigid mounting of air duct detector with sampling tubes.

most of the evolved gases in a fire are fuel specific and not associated with a sufficiently large number of fuels used for general-purpose detection. In large-scale fire testing where carbon monoxide and carbon dioxide measurements are generally taken, detectable levels of these gases occur between the occurrence of detectable particulate levels and detectable heat levels.

A gas-sensing fire detector can operate with simple sensors, such as a semiconductor or catalytic element, or more complex apparatus, such as non-dispersive infrared absorption. Fire-gas detectors with a semiconductor element respond to either oxidizing or reducing gases by creating electrical changes in the semiconductor. The subsequent change in conductivity of the semiconductor causes detector actuation. (See Figure 3-25.)

In a detector with a catalytic element, the detector contains a material that remains unchanged, but accelerates the oxidation of combustible gases. The resulting temperature rise of the element causes a change in the element's resistance, which then initiates the alarm.

A metal oxide semiconductor-type element is the operating element found in one type of gas-sensing fire detector. The element consists of an N-type semiconductor crystal in which two heater coils are helically wound and imbedded in opposite edges. The crystal is then coated with

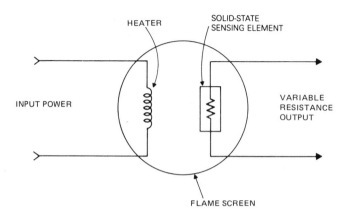

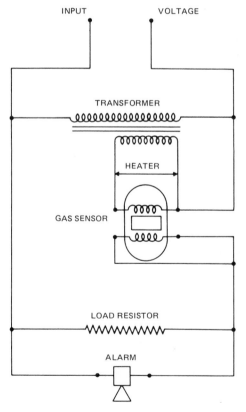

Figure 3-25. Gas-sensing fire detector with semiconductor element.

a metal (typically tin oxide) and supported by its heater wires within a metal flame-arresting mesh. One of the two heater wires is unpowered and serves as a single electrode. The other heater wire is operated at 5 volts dc to maintain the crystal at a temperature of about 662°F (350°C). The crystal must be operated at this temperature both to increase the number of free carriers available and also to burn off any contaminating gases that may come in contact with the crystal.

Under normal conditions, oxygen is adsorbed onto the surface of the crystal and gives it a certain conductivity. When any oxidizable gas comes in contact with the crystal, molecular oxygen is removed from the crystal surface and the conductivity increases. This increase is proportional to the gas concentration and triggers the alarm.

A disadvantage of many metal oxide elements is nonspecificity; i.e., the element will respond to any oxidizable gas, the response being additive in nature. This makes the device fairly prone to false alarms in areas where oxidizable gases are normally present in small amounts. However, the metal oxide semiconductor is well suited to applications where the combustible material involved is known to produce pyrolysis products that contain oxidizable gases. Recently, a CO-specific sensor has been developed and is being utilized in a CO detector for non-fire applications.

The catalytic hot-wire gas detector consists of a coil of platinum wire that is imbedded in an alumina bead and connected to terminal posts. (See Figure 3-26.) In a clear air condition, a specified low resistance is established when power is applied to the sensor element. When the sensor makes contact with a gas mixture, the molecules of the mixture oxidize on the sensing element, increasing the element's temperature and resistance.

To provide long-term stability, two elements are used in a Wheatstone bridge configuration. One element, the detector, has a catalytic coating. The compensating element, without a catalyst, does not respond appreciably to the presence of gas. In this configuration, power is supplied to the elements, heating them to their operating temperature. Resistors form the other side of the balanced bridge and establish a zero output signal.

When a gas mixture enters the detector housing, the mixture oxidizes on the catalytic coated element, increasing its temperature and resistance, unbalancing the bridge, and creating an output signal. A hot-wire bridge configuration compensates for changes in a detector due to temperature, aging, humidity, and power supply fluctuations. The hot-

CATALYTIC HOT-WIRE BRIDGE

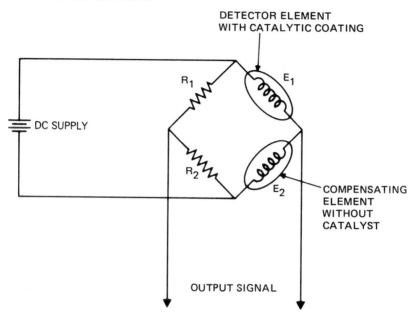

Figure 3-26. Gas-sensing fire detector with catalytic element.

wire sensor produces a linear output with respect to gas, creating a larger relative signal shift at the operating range than a semiconductor sensor.

A catalytic hot-wire gas detector has input power fed to a regulator, then onto the sensor bridge. (See Figure 3-27.) The output signal of the bridge is amplified, and the alarm horn is triggered and remains activated as long as the gas is present. A supervisory circuit monitors the sensing elements. Should an element fail, the bridge becomes completely unbalanced, sounding a trouble signal. An electrical test circuit is also connected to the bridge. Operation of the test switch alters the resistance of one part of the bridge, unbalancing the circuit in a way similar to the manner it is unbalanced in the presence of gas.

Another gas-sensing technique utilizes the change in color of a chemically treated material when it is exposed to a specific gas. Used since the 1940s for semi-quantitative gas measurement, the color change is typically nonreversable, and the sensors must be changed periodically.

Flame Detectors: Flame detectors respond to radiant energy that is either visible to the human eye (approximately 4,000 to 7,700 ang-

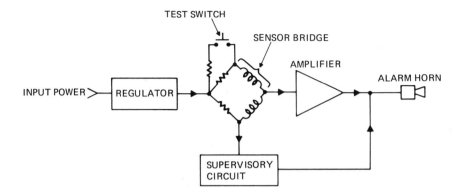

Figure 3-27. Catalytic hot-wire gas detector.

stroms) or outside the range of human vision. These detectors are sensitive to glowing embers, coals, or flames, which radiate energy of sufficient intensity and spectral quality to actuate the alarm.

Due to fast detection capabilities, flame detectors are generally used only in high-hazard areas, such as fuel loading platforms, industrial process areas, hyperbaric chambers, high ceiling areas, and atmospheres in which explosions or very rapid fires may occur. There are three general types of flame detectors: (1) infrared (IR), (2) ultraviolet (UV), and (3) combination IR/UV. Because flame detectors must be able to "see" the fire, they must not be blocked by objects placed in front of them. (The infrared type of flame detector, however, has some capability for detecting radiation reflected from walls.)

• Infrared-Type Flame Detectors: An infrared (IR) detector is basically a filter and lens system that screens out unwanted wavelengths and focuses the incoming energy on a photovoltaic or other type of cell that is sensitive to infrared energy. (See Figure 3-28.) IR flame detectors can respond to the total IR component of the flame alone, or to a combination of IR with flame flicker in the 5 to 30 Hz frequency range.

IR detectors that receive total IR radiation can be subject to interference from solar radiation, since the intensity of solar energy can be considerably larger than that of a small fire. This problem can be partially solved by using filters that exclude all IR not in the 2.5 to 2.8 micrometer and/or 4.2 to 4.5 micrometer ranges. These ranges represent absorption peaks for solar radiation due to the presence of CO_2 and

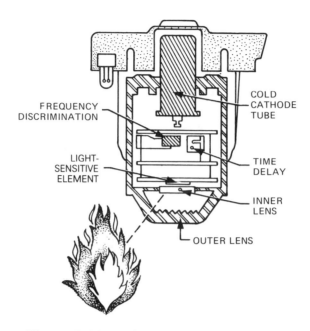

Figure 3-28.   Infrared-type flame detector.

water in the atmosphere. When IR detectors are used in locations that are shielded from the sun, such as vaults, this filtering is not necessary. Another approach to the solar interference problem is to employ two detection circuits, often referred to as a "two-color" system. One circuit is sensitive to solar radiation in the 0.6 to 1.0 micrometer range and is used to indicate the presence of sunlight. The second circuit will respond to wavelengths between 2 and 5 micrometers. A signal from the solar sensor circuit can be used to block the output from the fire-sensing cell, giving the detection unit the ability to discriminate against solar-based false alarms.

For most flame detector applications, flame flicker sensor circuits are preferred, since flicker and modulation characteristics of flaming combustion are not components of either solar or man-made interference sources. Use of flame flicker circuits results in an improved signal-to-noise ratio. These detectors use frequency-sensitive amplifiers with inputs tuned to respond to an alternating current signal in the flame flicker range. However, false alarms have been reported from sunlight reflecting off of the surface of water where waves produced the required flickering of the proper frequency.

Flame detectors are designed for volume supervision in either a fixed or scanning mode. The fixed units continuously observe a conical volume limited by the viewing angle of the lens system and the alarm threshold. The viewing angles range from 15 to 170 degrees for typical commercial units. One scanning device has a 400-ft (122-m) range and uses a mirror rotating at six revolutions per minute through 360 degrees horizontally, with a 100-degree viewing angle. The mirror stops when a signal is received. To screen out transients, the unit sounds an alarm only if the signal persists for 15 seconds.

Some flame detectors are designed to respond to passing sparks or flame fronts in ducts, such as those used in textile mills. These detectors do not contain a flame flicker circuit, but scan for glowing lint fibers in air ducting that might cause fires in the downstream filters. The detector turns on a water spray that extinguishes the glowing fiber before it reaches the filter.

• *Ultraviolet-Type Flame Detectors:* The ultraviolet (UV) component of flame radiation is also used for fire detection. (See Figure 3-29.) The sensing element of UV detectors can be either a solid-state device, such as silicon carbide or aluminum nitride, or a gas-filled tube in which the gas is ionized by UV radiation and becomes conductive, thus sounding the alarm. UV detectors have an operating wavelength range of 0.17 to 0.30 micrometers, and in that range are essentially insensitive to both sunlight and artificial light. UV detectors are also volume detectors, with viewing angles from less than 90 to 180 degrees.

• *Combination-Type Detectors:* The combination of UV and IR sensing has been applied to aircraft and hyperbaric chamber fire protection. These complex devices sound an alarm when there is a preset deviation from the prescribed ambient UV-IR discrimination level in conjunction with a signal from a continuous wire overheat detector. The analysis of the deviation is performed by an on-board minicomputer.

ALARM SIGNALS AND FIRE EXTINGUISHING SYSTEMS

Automatic fire extinguishing systems are often required to be installed in buildings that also have a fire alarm signaling system. When both exist it is often necessary (as required by local code) to connect the extinguishing system to the fire alarm system. The connection initiates a signal on discharge of the extinguishing agent and also provides supervisory alarm service of the extinguishing system. This supervisory ser-

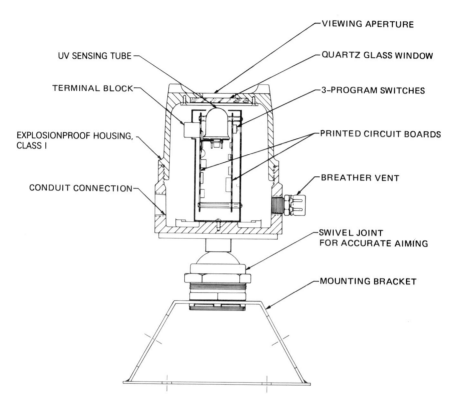

Figure 3-29. Ultraviolet-type flame detector.

vice indicates any off-normal condition of the system that would prevent normal operation and return of the system to a normal condition.

There are many types of extinguishing systems used with a fire alarm system; some of these are described in Table 3-3.

Functions of Alarm and Supervisory Signals

A sprinkler system with a waterflow alarm serves two functions: (1) that of an effective fire extinguishing system, and (2) an automatic fire alarm. Immediate notification by an alarm of the operation of sprinklers is important to the complete extinguishment of a fire and ultimate return of the system to service. Under some conditions, the sprinklers do not immediately or completely extinguish the fire; therefore, it is vital to have backup manual fire fighting forces notified to complete extinguishment either by portable extinguishing devices, private hose streams, or fire department equipment.

The amount of loss or damage by water after the fire has been extinguished can be held to a minimum by closing the control valve immediately after the need for sprinkler discharge has passed. One or two sprinklers may extinguish the fire, but water damage can be considerable unless the water is shut off as soon as it is safe to do so.

In addition to waterflow alarms, sprinkler systems are frequently equipped with devices to signal abnormal conditions that could make the protection inoperative or ineffective. In general, these supervisory devices give warning of troubles with equipment, e.g., shutoff valves, etc., and require action by maintenance or security personnel. Waterflow alarms and fire alarms alone give warning of the actual occurrence of a fire or other conditions, such as broken pipes, etc., causing water to flow through the system. Alarms alert occupants and summon the fire department. Any signal, whether waterflow or supervisory, can provide simply an audible local sprinkler alarm, or can be the initiating signal for a protective signaling system.

Waterflow Alarm and Supervisory Signal Service

Often a signaling system is installed in a building that has a sprinkler system. Provisions should be made to indicate the flow of water in a sprinkler system, and this indication should be made by an alarm signal within 90 seconds after flow of water begins equal to or greater than that from a single sprinkler of the smallest orifice size installed in the system. Movement of water due to waste, surges, or variable pressure need not be indicated.

Waterflow alarm system design for sprinklers utilizing on-off heads should ensure that an alarm will be received in the event of a waterflow condition. On-off sprinklers open at and close at predetermined temperatures. With certain types of fires, waterflow can occur in a series of 10- to 30-second short bursts. Waterflow detection devices with built-in time delays (called retards) may not detect waterflow under these conditions. It is recommended that an excess pressure system or one that operates on pressure drop be considered to facilitate waterflow detection on on-off sprinkler systems.

Excess pressure systems can be used with or without alarm valves. An excess pressure system with an alarm valve consists of an excess pressure pump with pressure switches to control pump operation. The inlet of the pump is connected to the supply side of the alarm valve, and the outlet is connected to the sprinkler system. The pump control pressure

TABLE 3-3. Fire Extinguishing Systems

Type	Description	Comments
1. Wet-pipe automatic sprinkler system.	A permanently piped water system under pressure, using heat-actuated sprinklers. When a fire occurs, the sprinklers exposed to the high heat open and discharge water individually to control or extinguish the fire.	Automatically detects and controls fire. Protects structure. May cause water damage to unprotected books, manuscripts, records, paintings, specimens, or other valuable objects. Not to be used in spaces subject to freezing. On-off types may limit water damage. See NFPA 13, *Standard for the Installation of Sprinkler Systems,* and NFPA 22, *Standard for Water Tanks for Private Fire Protection.*
2. Pre-action automatic sprinkler system.	A system employing automatic sprinklers attached to a piping system containing air that may or may not be under pressure, with a supplemental fire detection system installed in the same area as the sprinklers. Actuation of the fire detection system by a fire opens a valve that permits water to flow into the sprinkler system piping and to be discharged from any sprinklers that are opened by the heat from the fire.	Automatically detects and controls fire. May be installed in areas subject to freezing. Minimizes the accidental discharge of water due to mechanical damage to sprinkler heads or piping, and thus is useful for the protection of paintings, drawings, fabrics, manuscripts, specimens, and other valuable or irreplaceable articles that are susceptible to damage or destruction by water. See NFPA 13, *Standard for the Installation of Sprinkler Systems,* and NFPA 22, *Standard for Water Tanks for Private Fire Protection.*
3. On-off automatic sprinkler system.	A system similar to the pre-action system, except that the fire detector operation acts as an electrical interlock, causing the control valve to open at a predetermined temperature and close when normal temperature is restored. Should the fire rekindle after its initial control, the valve will reopen and water will again flow from the opened heads. The valve will continue to open and close in accordance with the temperature sensed by the fire detectors. Another type of on-off system is a standard wet-pipe system with on-off sprinkler heads. Here, each individual head has incorporated in it a temperature-sensitive device that causes the head to open at a predetermined temperature and close automatically when the temperature at the head is restored to normal.	In addition to the favorable feature of the automatic wet-pipe system, these systems have the ability to automatically stop the flow of water when no longer needed, thus eliminating unnecessary water damage. See NFPA 13, *Standard for the Installation of Sprinkler Systems,* and NFPA 22, *Standard for Water Tanks for Private Fire Protection.*
4. Dry-pipe automatic sprinkler system.	Has heat-operated sprinklers attached to a piping system containing air under pressure. When a sprinkler operates, the air pressure is reduced, a "dry-pipe" valve is opened by water pressure, and water flows to any opened sprinklers.	See No. 1. Can protect areas subject to freezing. Water supply must be in a heated area. See NFPA 13, *Standard for the Installation of Sprinkler Systems,* and NFPA 22, *Standard for Water Tanks for Private Fire Protection.*
5. Standpipe and hose system.	A piping system in a building to which hoses are connected for emergency use by building occupants or by the fire department.	A desirable complement to an automatic sprinkler system. Staff requires training to use hoses effectively. See NFPA 14, *Standard for the Installation of Standpipe and Hose Systems.*

TABLE 3-3. Fire Extinguishing Systems (continued)

Type	Description	Comments
6. Halon automatic system.	A permanently piped system using a limited stored supply of a halon gas under pressure, and discharge nozzles totally flood an enclosed area. Released automatically by a suitable detection system. Extinguishes fires by inhibiting the chemical reaction of fuel and oxygen.	No agent damage to unprotected books, manuscripts, records, paintings, or other irreplaceable valuable objects. No agent residue. Halon 1301 can be used with safeguards in normally occupied areas. Halon 1211 total flooding systems are prohibited in normally occupied areas. Halons may not extinguish deep-seated fires in ordinary solid combustibles, such as paper, fabrics, etc.; but are effective on surface fires in these materials. These systems require special precautions to avoid damage effects caused by their extremely rapid release. The high-velocity discharge from nozzles may be sufficient to dislodge substantial objects directly in the path. See NFPA 12A, *Standard on Halon 1301 Fire Extinguishing Systems*, and NFPA 12B, *Standard on Halon 1211 Fire Extinguishing Systems*.
7. Carbon dioxide automatic system.	Same as No. 6, except uses carbon dioxide gas. Extinguishes fires by reducing oxygen content of air below combustion support point.	Same as No. 6. Appropriate for service and utility areas. Personnel must evacuate before agent discharge to avoid suffocation. May not extinguish deep-seated fires in ordinary solid combustibles, such as paper, fabrics, etc.; but effective on surface fires in these materials. See NFPA 12, *Standard on Carbon Dioxide Extinguishing Systems*.
8. Dry chemical automatic system.	Same as No. 6, except uses a dry chemical powder. Usually released by mechanical thermal linkage. Effective for surface protection.	Should not be used in personnel-occupied areas. Leaves powdery deposit on all exposed surfaces. Requires cleanup. Excellent for service facilities having kitchen range hoods and ducts. May not extinguish deep-seated fires in ordinary solid combustibles, such as paper, fabrics, etc.; but effective on surface fires in these materials. See NFPA 17, *Standard for Dry Chemical Extinguishing Systems*.
9. High-expansion foam system.	A fixed extinguishing system that generates a foam agent for total flooding of confined spaces, and for volumetric displacement of vapor, heat, and smoke. Acts on the fire by: a. Preventing free movement of air. b. Reducing the oxygen concentration at the fire. c. Cooling.	Should not be used in occupied areas. The discharge of large amounts of high-expansion foam may inundate personnel, blocking vision, making hearing difficult, and creating some discomfort in breathing. Leaves residue and requires cleanup. High-expansion foam when used in conjunction with water sprinklers will provide more positive control and extinguishment than either extinguishment system used independently, when properly designed. See NFPA 11A, *Standard for Medium- and High-Expansion Foam Systems*.

switch is of the differential type, maintaining a constant sprinkler system pressure above the main pressure. Another switch monitors low sprinkler system pressure and initiates a trouble signal in the event of pump failure or other malfunction. An additional pressure switch can be used to stop pump operation in the event of a deficiency in water supply. Another pressure switch is connected to the alarm outlet of the alarm check valve to initiate a waterflow alarm signal when waterflow exists. This type of system also inherently prevents false alarms due to water surges. The sprinkler alarm retard chamber should be eliminated to enhance the detection capability of the system for short-duration flows. Off-normal conditions that could prevent normal operation of the system should also be monitored in sprinkler systems.

A dry-pipe sprinkler system equipped for waterflow alarm signals provides supplementary supervision of the system air pressure to avoid false signals due to neglect in maintaining air pressure. Signals transmitted should distinctively indicate the particular malfunction, e.g., valve position, temperature pressure, etc., of the automatic sprinkler system, plus its restoration to a normal condition.

Sprinkler system waterflow alarm and supervisory initiating devices and their circuits should be so designed and installed that they cannot be readily tampered with, opened, or removed from system connection without initiating a signal. This provision specifically includes junction boxes that have been installed on the outside of buildings to facilitate access to the initiating device circuit. No more than five waterflow switches should be connected to produce the same alarm signal, and no more than 20 supervisory switches connected to produce the same supervisory signal.

Monitoring of the various components of a sprinkler system supervisory signal service requires a distinctive signal indicating both the component off-normal condition and a different signal indicating return to normal condition. Generally, the required conditions essential to the proper operation of sprinkler systems should be supervised, except for those conditions related to water mains, tanks, cisterns, reservoirs, and other containers of water controlled by a municipality or a public utility.

Sprinkler system supervisory devices and equipment described in this chapter are considered to be a part of sprinkler installations even though they are manufactured, installed, and maintained by a central station or other supervisory service organization. Sprinkler system supervision is commonly provided for: (1) water supply control valves, (2) low water level in water supply tanks, (3) low temperature in water supply tanks or ground-level reservoirs, (4) high or low water level in pressure tanks, (5) high or low air pressure in pressure tanks, (6) high or low air pressure in dry-pipe sprinkler systems, (7) failure of electric power supply to fire pumps, (8) automatic operation of electric fire pumps, and (9) fire detection devices used in conjunction with deluge and/or preaction and recycling systems. Sprinkler system devices that give waterflow alarms or supervise the condition of the installation are shown schematically in Figure 3-30.

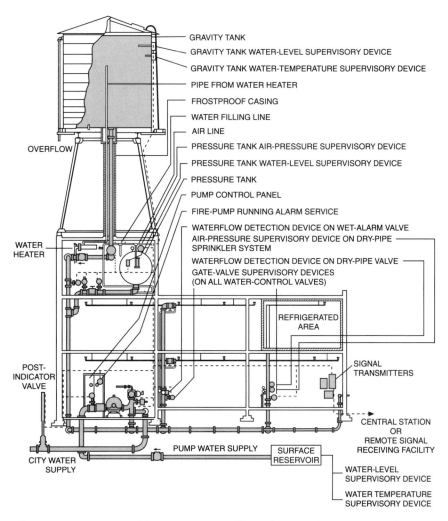

Figure 3-30. Sprinkler system waterflow alarm and supervisory devices.

An extinguishing system control valve should be supervised if movement of the valve from its normal position restricts or prevents the proper operation of the system. Two separate and distinct signals should be obtained, one indicating movement of the valve from its normal position and the other indicating restoration of the valve to its normal position. All supervisory parts of a system, i.e., water pressure, level, and temperature, should have two separate and distinct signals as well. The off-normal signal should be obtained during the first two revolutions of the handwheel or during one-fifth of the travel distance of the valve control apparatus from its normal position.

The off-normal signal should not be restored at any valve position except normal. An attachment for supervising the position of a control valve must not interfere with the operation of the valve nor obstruct the view of its indicator nor prevent access to its stuffing box.

Where the signaling attachments of two or more valves utilize a common circuit, a restoration signal should be obtained only when all valves in the group are in their normal positions.

Pressure Supervision: A pressure supervisory signal attachment for a pressure tank and one for a dry-pipe sprinkler system should indicate both high- and low-pressure conditions. A signal should be obtained when the required pressure in the tank is increased or decreased 10 psi (70 kPa) from the required pressure value. Pressure change signals for a dry-pipe system should be used in accordance with the requirements of the authority having jurisdiction where the system is used.

A steam-pressure supervisory attachment should indicate a low-pressure condition, and a signal obtained when the required pressure is reduced to a value not less than 110 percent of the minimum operating pressure of the steam-operated equipment supplied. An attachment for supervising the pressure of sources other than those previously specified should be capable of being applied and operated as required by the local authority.

Water Level Supervision: A pressure tank supervisory attachment should indicate both high- and low-level conditions in water storage. A signal should be obtained when the water level is lowered or raised 3 in. (76 mm) from the required level. A supervisory attachment for other than pressure tanks should indicate a low-level condition and a signal obtained when the water level is lowered 12 in. (300 mm) from the required level.

Water Temperature Supervision: For exposed water storage containers, one separate and distinct signal indicates that the temperature of the water has been lowered to 40°F (4.4°C), and the other indicates restoration to the required temperature.

Pump Supervision: Automatic fire pumps and special fire service pumps should be supervised as the authority having jurisdiction may require. Where supervision is applied to the electric power supplying the pump, the supervisory device should be connected on the line side of the motor starter so open fuses or open circuit breakers in the supply line to the pump will be detected promptly. All phases of operation should be supervised. All fire pump supervisory signals should be independent of, or not affected by, other signals except that engine drive high temperature, low oil, overspeed shutdown, battery failure, and failure to start may have a common signal.

The items that can be supervised in pump operation include, but are not limited to: electric drive (pump running, power failure, phase reversal), engine drive (pump running, controller not in "auto" condition, high engine temperature, low lubricating oil pressure, failure to start, overspeed shutdown, battery failure), and steam drive (pump running if practical, low steam pressure supervision). NFPA 20, *Standard for the Installation of Centrifugal Fire Pumps*, provides more information on pump operation.

Waterflow Sprinkler Alarms: The various types of sprinkler alarms include: (1) those that operate with an actual flow of water; (2) those that activate a hydraulic or an electric alarm when a water control device such as a dry-pipe valve trips to admit water to the alarm device or mechanically operates an electric switch, whether or not water actually flows from sprinklers; and (3) those that not only signal the tripping of the control valve but may also give supplementary warning signals either in case of damage that might impair the operation of the system, or if maintenance features need attention.

Sprinkler systems are usually required to have an approved water motor gong or an electric bell, horn, or siren on the outside of the building. An electric bell or other audible signal device may also be located inside the building. Water-operated devices should be located near the alarm valve, dry-pipe valve, or other water control valves in order to avoid long runs of connecting pipe.

All electric alarm devices, wiring, and power supplies should comply with the appropriate local codes.

Wet Pipe Sprinkler System Alarm Devices: Waterflow alarm devices have been used to some extent ever since automatic sprinklers were first installed. They are generally located at or near the base of sprinkler risers but may be used as floor or branch alarms. They are designed and adjusted to give an alarm if a waterflow equal to the discharge of one or more automatic sprinklers occurs in the sprinkler system. The alarm signal can be given electrically, by water motor gongs, or by both. By far the most common types of waterflow alarm devices are the waterflow alarm valve and the waterflow indicator. (See Figure 3-31.)

The basic design of most waterflow alarm valves is that of a check valve which lifts from its seat when water flows into a sprinkler system. The movement of the valve clapper is used in either of the following ways:

1. The valve seat ring can have a concentric groove with a pipe connection from the groove to the alarm devices. Such valves are commonly called differential-type (Figure 3-32, top) or divided-seat ring-type valves. When the clapper of the alarm valve rises to allow water to flow to sprinklers, water also enters the groove in the divided-seat ring and flows through the pipe connection to an alarm-signaling device.
2. The clapper of the alarm check may have an extension arm connected to a small auxiliary (pilot) valve having its own seat and a pipe connection to alarms. (See Figure 3-32, bottom.) When the auxiliary valve is lifted by movement of the main valve clapper, water is admitted to alarm devices.

Waterflow Indicators: A waterflow indicator of the paddle or vane type consists of a movable flexible vane of thin metal or plastic that is inserted through a circular opening cut in the wall of a sprinkler supply pipe. The vane extends into the waterway sufficiently to be deflected by any movement of water flowing to opened sprinklers. (See Figure 3-33.) Motion of the vane operates an alarm-actuating electric switch or mechanically trips a signaling system transmitter. A mechanical, pneumatic, or electrical time delay feature in the detector or electric circuit, or a part of the signaling system transmitter, prevents false alarms being given by fluctuating water pressures. The retard feature should be of the

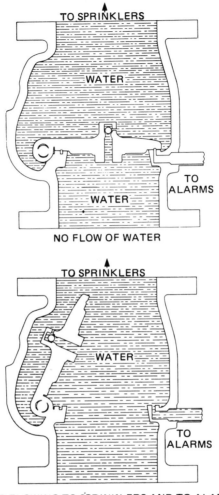

TO SPRINKLERS

WATER

WATER

TO ALARMS

NO FLOW OF WATER

TO SPRINKLERS

WATER

TO ALARMS

WATER FLOWING TO SPRINKLERS AND TO ALARMS

Figure 3-31. A wet-pipe sprinkler system is under water pressure at all times so water will discharge immediately when automatic sprinklers operate. Automatic alarm valve shown causes a warning signal to sound when water flows through the sprinkler piping.

instantly recycling type or otherwise arranged so that the effect of a sequence of flows, each of less duration than the predetermined retard period, will not have a cumulative effect. Electrically heated thermal retards have not performed satisfactorily. Waterflow indicators have no provision for supplying water to water motor alarm gongs.

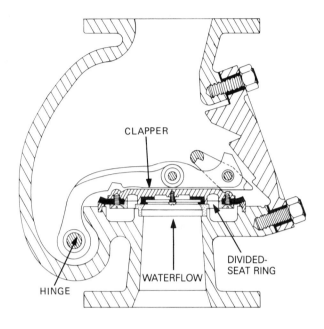

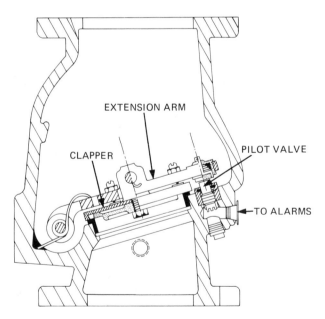

Figure 3-32. (Top) Differential-type waterflow alarm valve; (Bottom) alarm check valve with auxiliary valve and pipe connection to alarms.

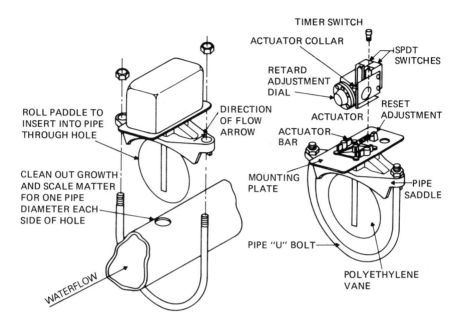

Figure 3-33. Vane-type waterflow indicator switch.

It is important that the flexible vane of a waterflow indicator be of a design and material not subject to mechanical injury or corrosion so that it cannot become detached and possibly obstruct the sprinkler piping. Waterflow indicators of the vane type cannot be used in dry-pipe systems, deluge systems, or preaction systems, because the vane and mechanism are likely to be damaged by the sudden rush of water when the control valve opens.

Waterflow indicators are commonly used on new sprinkler systems. Situations where their use is most prevalent are where: (1) ease of installation and economy are important in installing a waterflow alarm, (2) subdivision into several alarm areas is desired on a large sprinkler system, or (3) a central station, proprietary system, or other remote signal service is available that can receive only electric alarm signals from the protected property. Waterflow indicators are supplied by manufacturers of both sprinkler equipment and signaling system equipment.

Alarm Retarding Devices: An alarm check valve that is subjected to fluctuating water supply pressures needs an alarm retarding device in order to prevent false alarms when the check valve clapper is lifted from its seat by a transient surge of increasing pressure.

Retarding Chambers: One type of device, usually designated as a retarding chamber, is essentially a chamber inserted in the water line from the alarm check valve to the water motor gong and electric circuit closer. (See Figure 3-34.)

Flows of short duration from the alarm check valve to the alarm device first accumulate in the retarding chamber which must become filled before water passes to the alarm device. The size of the chamber and the water inlet are predetermined by the manufacturer to give a delay needed to ensure continuous waterflow before an alarm is given. The air displaced from the alarm piping in advance of water escapes at low pressure and does not operate the alarm devices.

The retarding chamber is self-draining so that, unless surges follow in close succession, the full retard interval is restored between surges.

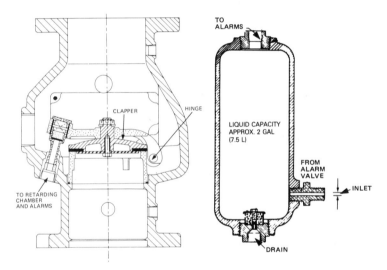

Figure 3-34. (Left) Waterflow alarm valve; and (Right) retarding chamber.

Pressure-Actuated Alarm Switches

Electric switches, frequently called circuit closers, have contacts arranged to open or close an electric circuit when subjected to increased or reduced pressure. These switches are used in combination with dry-pipe valves, alarm check valves, and other types of water control valves to initiate an electric waterflow alarm signal when a flow of water to sprinklers occurs. Electric switches can also give a supervisory signal if pressures increase or decrease beyond established limits.

Manufacturers of dry-pipe valves, alarm check valves, and special types of water control valves regularly furnish approved pressure-operated switches to operate local electric waterflow alarms. In most cases, the motion to actuate a switch is obtained from a diaphragm exposed to the pressure on one side and opposed by a fixed adjustable spring on the other side. A typical waterflow alarm switch is shown in Figure 3-35.

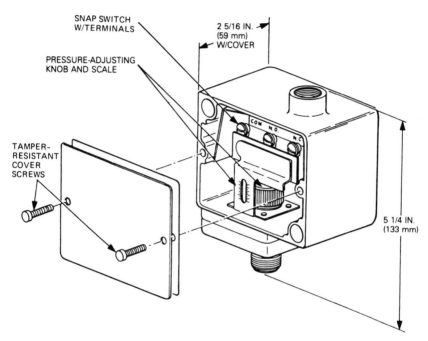

Figure 3-35. Isometric view of waterflow alarm switch. Water pressure from the system acts upon a diaphragm in the housing at the base of the switch enclosure, causing movable contacts to close with stationary contacts. The switch can be used for normally closed or open circuits, and sensitivity to pressure differentials adjusted from 5 to 15 psi (34.0 to 103.0 kPa).

Mercury switches and other types of electric contacts can be used. Commercial pressure-actuated switches can be adapted to waterflow alarm service.

Other Supervisory Devices

General requirements for supervisory signaling systems include electrical supervision of circuits to indicate conditions that could prevent the

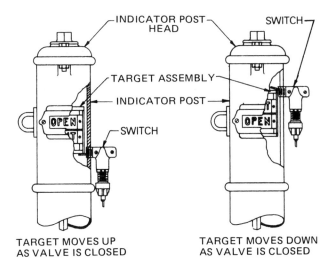

TARGET MOVES UP
AS VALVE IS CLOSED

TARGET MOVES DOWN
AS VALVE IS CLOSED

Figure 3-36. An indicator post valve supervisory switch. The switch operating stem is held against the movable target assembly by springs within the switch housing. The switch end of the operating stem carries an insulator that separates contact springs and opens the electrical circuit when the target assembly is moved in the valve closing direction by approximately two turns of the valve stem.

required operation of the sprinkler system. The most common condition is a shut control valve. Electric switches for supervision of sprinkler system water supply control valves are of different mechanical designs for the different types of control valves. Electric switches can be for open- or closed-circuit supervisory systems. Supervision can also be by means of signaling transmitters mechanically tripped by operation of the gate valve.

The signal to indicate valve operation is given within two turns of the valve wheel from the wide open position. The restoration signal is given when the valve is restored to its fully open position. Figures 3-36 and 3-37 show representative approved devices for supervision of sprinkler valves.

The temperature of the water in fire service tanks exposed to cold weather is usually shown by a thermometer in the cold water return to a circulating-type heater or near the bottom of a large riser. Supervisory equipment is available for detecting dangerously low temperature near the surface of the water at the tank shell where freezing is most likely to start, as well as to check manual supervision or automatic heat.

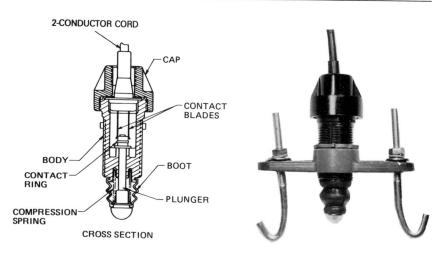

Figure 3-37. A gate valve supervisory switch. Hook bolts (photo at right) hold the switch to the two sides of the valve yoke. When the valve is wide open, the plunger tip enters a ⅛-in. (3-mm) deep depression drilled in the valve stem. The switch is adjusted so that, in the open valve position, the contact ring closes the electric circuit between the contact blades. When the valve is closed not over two turns, the plunger tip rides out of the depression in the valve stem, opening the electric circuit. If the switch is tampered with or removed, the plunger takes the position shown in the sectional view (diagram at left) and opens the electric circuit.

Devices for supervising the water level in pressure tanks differ from those used in water supply tanks, in that level-sensing elements are continuously subject to high pressure. The supervisory signal is given if the water level reaches 3 in. (76 mm) above or below the proper point, usually by means of an electrical connection to a separate signal transmitter. Figure 3-38 shows a water-level supervisory switch for gravity tanks.

Guard's Tour Supervisory Service

During normal business hours, most areas of most facilities are occupied. By their presence, the occupants can provide fire protection surveillance if they are in the area when a fire occurs. A carelessly discarded cigarette, for example, may ignite the contents of a wastebasket; that fire may spread until the entire building is involved. However, someone who discovers the fire while it is still in its incipient stage may be able to extinguish it with a portable fire extinguisher. In some cases, the occupants can also detect conditions such as a malfunctioning

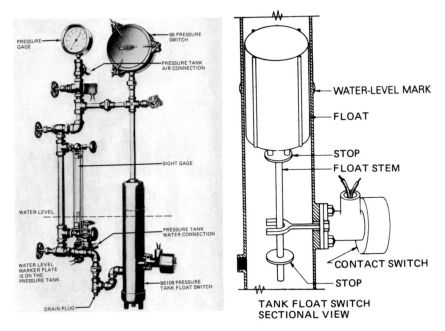

Figure 3-38. A supervisory switch for pressure water tanks. If the water level rises or falls about 3 in. (76 mm), one of the stops on the float stem in the tank float switch comes in contact with the forked lever which operates the contact switch. The switch is usually connected into the circuit of a supervisory signal transmitter.

machine that might lead to a fire. Further, most companies have come to realize that increased security surveillance is vital in guarding against fires of incendiary origin.

Guard services generally serve three purposes to protect a property against fire loss: (1) to protect the property at times when the management is not present; (2) to facilitate and control the movement of persons into, out of, and within the property; and (3) to carry out procedures for the orderly conduct of some operations on the property. Guards may be facility employees or employees of outside firms established to provide these services on a contract basis. The duties of these individuals can be supplemented or replaced in part by various approved protective signaling systems.

Guard supervisory services designed to continuously report the performance of a guard are found in connection with local signaling service, central station service, and proprietary protective signaling systems. These services usually provide for supervised or compulsory tours.

For supervised tours, a series of patrol stations along the guard's intended route are successively operated by the guard with each station sounding a distinctive signal at a central headquarters. Customarily, the guard is expected to reach each of these stations at a definite time, and failure to do so within a reasonable grace period prompts the central station to investigate the guard's failure to signal. Frequently, manual fire alarm boxes that ordinarily transmit four or five rounds of signals for fire can also be actuated by a special watch key carried by the guard to transmit only a single round to the central station, thus signaling that the box has been visited.

By proper location of the stations, a fire or security guard can be compelled to take a definite route though the premises, and variations from that route would appear as misplaced signals on the recording tape. The order of station operation can be modified on occasion in the interests of security or to meet special conditions within the building.

With compulsory tours, one or more stations are wired to the central station. Preliminary mechanical stations condition the guard's key to operate the wired station only after the preliminary stations have been operated in a prearranged order. This compulsory tour arrangement is somewhat less flexible than supervised tours but has the advantages of the absence of interconnected wires between the preliminary stations and the reduction of signal traffic. The usual arrangement is to have the guard transmit only start and finish signals that must be received at the central point at programmed reception times.

Specific features are required for a guard's tour system that is provided as part of a fire alarm signaling system. NFPA 601, *Standard on Guard Service in Fire Loss Prevention*, should be consulted concerning the number of guard's reporting stations, their locations, and the route to be followed by the guard for operating the stations.

A permanent record indicating each time each signal-transmitting station is operated should be made at the main control unit. When intermediate stations that do not transmit a signal are employed in conjunction with signal-transmitting stations, distinctive signals should be transmitted at the beginning and end of each guard's tour and a signal-transmitting station provided at intervals not exceeding ten stations. Intermediate stations that do not transmit a signal should be capable of operation only in a fixed order of succession.

In addition to these general provisions, the system should transmit a "start" signal to the main control unit. The guard should initiate this signal at the start of continuous tour rounds. A "delinquency" signal-should be automatically transmitted within 15 minutes after a predetermined time if the guard fails to actuate tour stations as scheduled. A "finish" signal is transmitted on a predetermined schedule after the guard completes the last tour of the premises. A "start" signal should be transmitted at least once every 24 hours for periods of over 24 hours, during which tours are continuously conducted. The "start," "delinquency," and "finish" signals should be recorded at the main control unit.

BIBLIOGRAPHY

NFPA Codes, Standards, Recommended Practices, and Manuals. (See the latest *NFPA Catalog* for availability of current editions of the following documents.)

NFPA 20, *Standard for the Installation of Centrifugal Fire Pumps.*

NFPA 70, *National Electrical Code.*

NFPA 72, *National Fire Alarm Code.*

NFPA 110, *Emergency Power Systems.*

NFPA 601, *Standard on Guard Service in Fire Loss Prevention.*

4

Signal Transmission

INTRODUCTION

This chapter discusses three types of fire alarm system signaling circuits (channels): (1) wire, (2) wireless, and (3) optical fiber. The circuits connect alarm initiating devices to the fire alarm control panel and other locations that require notification of a fire. Methods of transmission, compatibility of initiating devices, signal verification, and false alarms are also covered.

A copper wire is the oldest method of carrying a signal from one component of a fire alarm system to another. The physical and electrical use of this type of signaling circuit is more completely described in Article 760 of NFPA 70, *National Electrical Code.*

Wireless transmission is used to transmit signals in proprietary, central station, remote station, and public fire service communication systems. Wireless transmission also transmits alarm signals from initiating devices, such as smoke detectors, manual fire alarm boxes, etc., to the fire alarm control unit.

Requirements for optical fiber signaling circuits are more fully described in Article 770 of NFPA 70, *National Electrical Code.* For related information see Chapter 5, "Signal Processing"; Chapter 7, "Signal Notification"; and Chapter 13, "Installation."

WIRE TRANSMISSION

Conventional Detectors

Transmitting alarm signals by wire can be as basic as increasing the current in the circuit by shunting the end-of-line resistor when a switch or

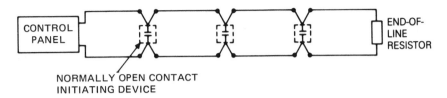

Figure 4-1. Basic wired alarm initiating circuit.

relay contact closes in a manual fire alarm box, heat detector, or smoke detector. (See Figure 4-1.)

When smoke detectors are used on the type of alarm initiating circuit shown in Figure 4-1, separate conductors can be provided to power the smoke detectors. (See Figure 4-2.) These detectors, called 4-wire detectors, can have contacts in addition to the relay alarm contacts to perform supplementary or releasing service functions. Since the smoke detector electronics and relay are not dependent on the limited power available from an alarm initiating circuit, more than one smoke detector alarm relay can be energized simultaneously on the circuit.

A power supervisory relay can be installed electrically beyond the last detector on the circuit so the detector power source is properly supervised. Loss of power or disconnect of any field power wire causes the relay to deenergize, its contact to open, and a trouble condition to be indicated at the control panel.

The initiating circuit current can be increased by shunting the end-of-line resistor electronically, as in a smoke detector (a "2-wire" detector) that receives power and transmits an alarm signal over the same pair of wires. Field-wiring of 2-wire detectors is shown in Figure 4-3.

Two-wire detectors should be used carefully. The detectors should be electronically compatible with each other and the fire alarm control panel. If relays in the detectors are intended to perform supplementary functions, most likely the first relay on that circuit would be the only one to operate unless the circuits are arranged to ensure operating power.

Relays mounted in 2-wire detectors should not be used to control essential functions of the fire alarm system because operation of the supplemental relay depends on the limited power available on the initiating circuit. After the first smoke detector senses smoke and the relay energizes, there is usually insufficient power available to energize a relay in a second detector on that circuit. Even though it normally requires less current to maintain an energized relay than to initially

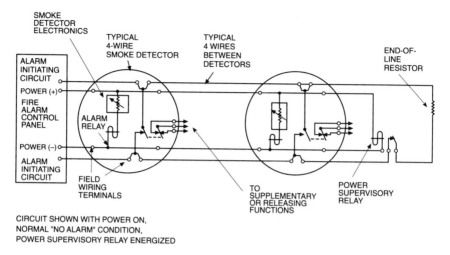

Figure 4-2. Typical 4-wire smoke detector initiating circuit.

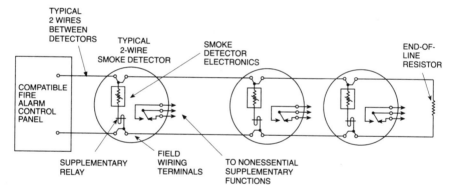

Figure 4-3. Typical 2-wire smoke detector alarm initiating circuit.

energize the relay, if other initiating devices are operating and placed in parallel on the circuits, the first relay could be deprived of sufficient current, causing it to deenergize. Operation of a contact-type device, such as a heat detector or manual fire alarm box, on the same 2-wire circuit would render all smoke detectors and built-in relays inoperative.

Addressable Detectors

The term "addressable" applies to various types of signal transmission between a detector and the fire alarm control panel. The following examples illustrate some of the various methods for addressable signal transmission.

1. The detector can wait to be addressed at each scanned (polled) cycle of the control panel and respond with a digital signal to indicate an alarm or normal condition. If no response is received at the control panel, a trouble (fault) condition is indicated.

2. An analog output signal that is processed at the control panel determines the moment at which the signal should be considered a fire alarm. In this example, a pre-alarm is indicated at the control panel when the signal reaches a certain point below the programmed alarm threshold.

3. An individual detector on a circuit can be identified by a time-polling sequence, in which the status of the first detector only on the circuit is shown when the panel polls the circuit. After a programmed time period, each detector switches the polling signal to each succeeding detector on the line. The individual detector can be identified at the fire alarm panel when the time slot for detector response is shown.

4. The fire alarm control panel can monitor the analog signal from a detector and determine validity of a fire alarm signal based on the rate of change of the signal. For example, a rapid increase in the analog signals for a few scans, or a slow increase for a period of many scans, can be used to signal a fire condition.

5. A specific number of control panel scans can be counted to determine that a true alarm condition exists instead of a transient condition appearing in one or two scans.

6. A particular identification response signal can be plotted on a display at the control panel, pinpointing the location of the detector.

Addressable detectors require clear specifications that should be developed prior to purchasing a system, since each method requires its own specific design features. For example, fewer wires are needed to connect addressable detectors to the control panel, or more information on the status of each detector must be obtained.

Multiplexing

Multiplexing is a signaling method characterized by simultaneous or sequential transmission, or both, and reception of multiple signals in a communication channel. Multiplexing includes the means for positively identifying each signal. Multiplexing of signals has been used for central station, proprietary, and public fire service communication systems for many years.

The earliest form of signal system multiplexing was the use of a coded system. Addressable detectors are a more recent form of multiplexing.

Many traditional coded systems are still in use; however, modern multiplexing uses either a microprocessor or a computer. There are two multiplexing methods: (1) active and (2) passive.

Active Multiplexing: Active multiplexing is defined as the transmission of status signals from an initiating device circuit or an initiating device to a central receiving station upon command, enabling individual identification of signals from many locations.

In an active multiplex system, a transponder or circuit usually interfaces one or more initiating device circuits with a signaling line circuit in a manner that permits individual identification and display of each indicating circuit status at a central supervisory station.

• Microprocessor Control Units: Microprocessor control units in an active multiplex system consist of thousands of transistors incorporated into an integrated circuit. The microprocessor is a single device capable of performing many basic operations of a digital computer and is highly cost effective in labor and wire savings. In designing very large fire alarm signaling systems, a designer should decide whether to incorporate microprocessor-based equipment and multiplex the information throughout the building, or to provide a hard-wired system with a standard solid-state control panel.

Some microprocessor multiplex systems are specified with integral remote multiplexing for water and waste treatment plants, plant monitor and control systems, building automation systems, and fire alarm systems.

Microprocessor-based control equipment offers many advantages not available with standard solid-state equipment. For example, microprocessors can include system diagnostics, allowing the contractor and/or maintenance personnel to view system information and determine problems with a particular control panel.

The microprocessor control unit is the basis for an integrated fire protection system that incorporates all connected, addressable devices, as well as all internal modules to ensure their proper operation. The microprocessor design allows the sensitivity of each smoke detector to be analyzed to determine alarm, normal, and trouble conditions. The interconnection line can be a single pair or noncontinuous (as in a power distribution system), i.e., spider-webbed or T-tapped, respectively. (See Figure 4-4.)

Most microprocessor-based control units have been developed for use with complex multiplex systems. Systems sold today have modular design, which allows the system capacity to match the initial design

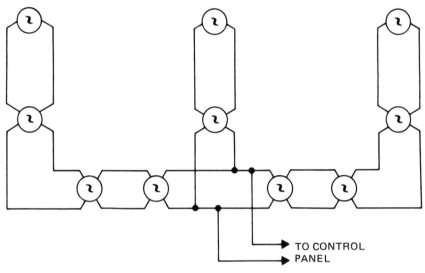

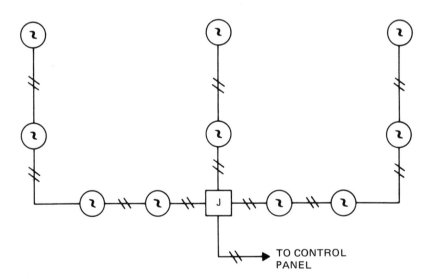

Figure 4-4.   Spider-webbed/T-tapped initiating circuit riser diagram.

requirements and allows for system expansion. Microprocessors, however, are complex and must be serviced by qualified personnel.

The microprocessor central processing unit can automatically operate the system control points as a result of an alarm condition from any of its monitoring and sensoring points.

A microprocessor-based system will provide much information; much of which may be unnecessary for the fire service, but may assist the owner to more efficiently operate the building complex.

The owner/operator personnel and the fire department personnel who will be responsible for fire fighting operations in the building should all receive adequate training on the system from the manufacturer.

• *Multiplexing Formats for Two-Way Communication:* These formats can be one of three types: (1) ripple through, (2) sequential counting, or (3) digitally addressable.

The ripple-through format is a response to a "start" command and each device in turn transmits its status. (See Figure 4-5.) One form closes the path to the following device upon completion of transmission from the responding device. Devices are usually connected in series to the transmission pathway, which is commonly a 2-wire circuit. These devices require no individual address, since the address is their position in the sequential arrangement. Spider-webbed or T-tapped distribution is not possible with this format.

For the sequential counting format, station No. 1 reports in response to a synchronizing (reset) signal and the first transmit command. After the second transmit command, station No. 2 reports and this continues until all stations have responded. (See Figure 4-6.) Spider-webbed or T-tapped distribution is inherent in this format and end-of-line devices are not required. Emergency operation, i.e., operation with a single open, is easily achievable. An individual address setting at each remote device is required, however, and interrupt capability is not possible.

For the digitally addressable format, each station has a unique digital address, making a random order communication possible. (See Figure 4-7.) When the response includes the address/location of the station, interrupt capability becomes possible. This format provides maximum functional flexibility as well as interrupt capability, but also requires the largest number of characters to communicate. This slows down the polling cycle and necessitates higher speed transmission.

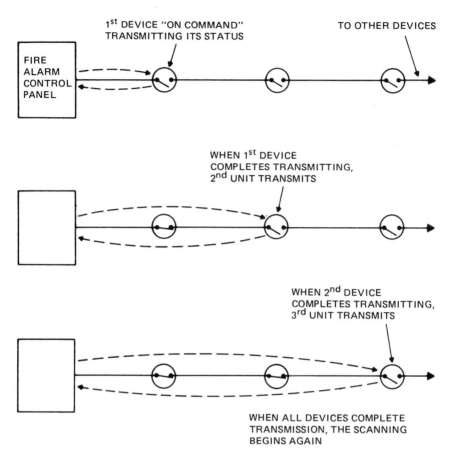

Figure 4-5.   Ripple-through multiplex transmission.

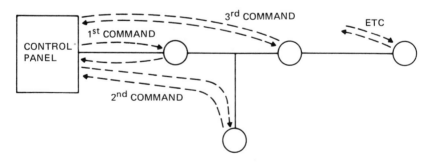

Figure 4-6.   Sequential counting multiplex transmission.

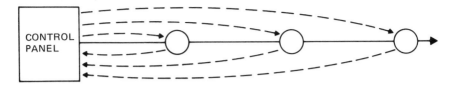

Figure 4-7. Digitally addressable random multiplex transmission.

• *The Concept of Networking:* When a number of addressable detectors are connected to a multiplexed initiating device loop, the central control point of the loop may in turn act as a transponder in a system's active multiplex circuit. (See Figure 4-8.) The detailed information gathered from the multiplexed initiating device loop can be arranged for retransmission to the system supervisory control station.

Passive Multiplexing: Passive multiplexing removes the command to transmit and permits transmitters to report at any time. The system becomes a one-way system with multiple transmitters sharing the same communication channel or pathway.

A simple coded fire alarm system is a passive multiplex system. More recently, battery-powered radio frequency (RF) transmitters have been introduced in passive multiplex systems and can identify themselves and transmit this information when an alarm condition occurs.

Analog Data Transmission

Conventional spot-type detectors have two stable states: (1) normal and (2) alarm. The detector changes from the normal to the alarm state when the signal produced by the environment is compared in the detector to a reference level determined by the sensitivity calibration of the detector. In more sophisticated detectors there may be multiple reference levels allowing a detector to indicate a pre-alarm condition.

In a sense, a detector that transmits a signal indicative of the signal produced by the environment is an analog detector that simply repeats the signal from its sensor in a form compatible with the data transmission format.

Analog detectors combined with addressable devices provide a system where a detector simply reports the signal produced by its environment to the control equipment, and the decision to sound an alarm or other response is produced by the control equipment.

With analog data transmission it is possible to compensate for long-term changes in sensor response and essentially maintain a constant

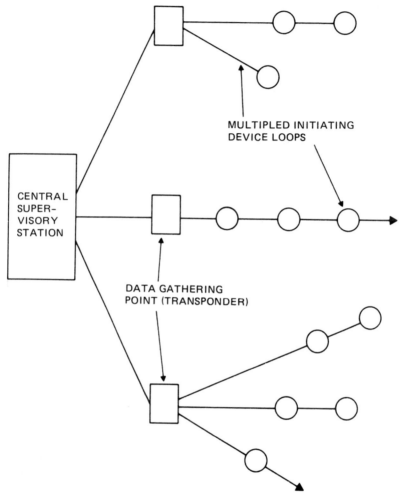

Figure 4-8. Networking, active multiplexing circuit.

sensitivity. The apparent alarm threshold can be easily changed to accommodate cyclical changes, i.e., day to night, in the environment; and to protect difficult hazards.

WIRELESS TRANSMISSION

A wireless initiating device is defined as any initiating device that communicates with associated control/receiving equipment by some kind of wireless transmission medium. Requirements for the use of wireless, i.e., radio frequency or "RF," transmission vary for each type of protective signaling system.

Central Station Systems

Wireless transmission for central station systems is accomplished by two-way RF multiplex systems. Central station equipment, transponders, transmitters, and receivers should comply with Federal Communications Commission (FCC) rules and regulations. Wiring, power supplies, and overcurrent protection requirements can be found in NFPA 70, *National Electrical Code.*

RF multiplex systems should be installed in accordance with NFPA 70, *National Electrical Code.* All external antennae should be protected in order to minimize the possibility of damage by static discharge or lightning.

If duplicate equipment for signal receiving, processing, display, and recording is not provided, the installed equipment should be so designed that any critical assembly can be replaced from on-premises spare parts and the system restored to service within 30 minutes. A malfunction in a critical assembly is one that will prevent the receipt and interpretation of signals by the central station operator. If RF multiplex equipment fails, spare modules, such as printed circuit boards, printers, etc., should be stocked at the central station to expedite repairs.

Operation of Signal Transmission: Any status changes that occur in an initiating device or in the interconnecting communication channel from the location of the initiating device(s) to the central station should be presented in a form the operator can interpret quickly. Status change signals should provide information on type of signal, i.e., alarm, supervisory, delinquency, or trouble signal; condition, i.e., differentiate between an initiation of an alarm, supervisory, delinquency, or trouble signal, and restore or return to normal; and location, i.e., point of origin.

Many methods of recording and display or indicating change-of-status signals are possible. The following are recommended:

1. Each change-of-status signal requiring operator action should result in an audible signal, and a minimum of two independent methods should identify the type, condition, and location of the status change.
2. Each change-of-status signal should be automatically recorded with type of signal, condition, and location in addition to the time and date the signal was received.

3. Failure of an operator to acknowledge or act upon a change-of-status signal should not prevent subsequent alarm signals from being received, indicated or displayed, and recorded.

4. Change-of-status signals requiring operator action should be displayed or indicated in a manner that clearly differentiates these signals from those that have been acted upon and acknowledged.

There are maximum operating times or maximum end-to-end operating time parameters recommended for a two-way RF multiplex system. Maximum allowable time lapse from initiation of a single fire alarm signal until it is recorded at the central station should not exceed 90 seconds. Subsequent fire alarm signals should be recorded at a rate no slower than one every additional 10 seconds. The maximum allowable time lapse from the occurrence of an adverse condition in any interconnecting communication channel (until recording of the adverse condition is started) should not exceed 90 seconds.

In addition to the maximum operating time above for fire alarm signals, the following are suggested: systems with more than 500 initiating device circuits should be able to record a minimum of 50 simultaneous status changes in 90 seconds. Systems with fewer than 500 initiating device circuits should be able to record a minimum of 10 percent of that total number of simultaneous status changes within 90 seconds.

Central station, satellite, and repeater station radio transmitting and receiving equipment should be controlled and supervised from the central station. This can be accomplished via a supervised circuit in which the radio equipment is remotely located from the central station.

Transmitter in use (radiating), failure of ac power supplying the radio equipment, receiver malfunction, and indication of automatic switchover should be supervised at the central station. Independent deactivation of transmitters should be controlled from the central station.

RF Communication Channel: The RF multiplex communication channel terminates in a transponder at the protected premises and in a system unit at the central station. Channels can be private facilities, such as microwave, or leased facilities furnished by a communication utility company.

Occurrence of an adverse condition on the channel between a protected premises and the central station that will prevent the transmission of any change-of-status signal should be automatically indicated and recorded at the central station. This identifies the affected portions of the system so the central station operator can determine the location of the adverse condition by trunk and/or leg facility. Restoration of normal service to the affected portions of the system should be automatically recorded. The first status change of any initiating device circuit that occurred at any of the affected premises during the service interruption should also be recorded.

Dual control should provide for redundancy between a receiving station and the central station in the form of a standby channel or similar alternative means of transmitting signals over the primary trunk portion of a communication channel. The same method of signal transmission can be used over separate routes, or different methods of signal transmission can be utilized. Public switched telephone network facilities should be used only as an alternate method of transmitting signals.

When leased telephone lines are used, that portion of the primary trunk between the central station and its serving wire center may be excepted from the separate routing requirement of the primary trunk. Where used, dual control should be supervised as follows.

Dedicated facilities that are available full time with limited use for signaling purposes should be exercised at least once every hour. Public switched telephone network facilities should be exercised at least once every 24 hours.

Radio multiplex systems are classified based upon their ability to perform under adverse conditions that affect their communication channels. System classifications are of two types: (1) Type 4 and (2) Type 5.

A Type 4 system has two or more control sites. Each site has a receiver interconnected to the central monitoring point by a separate communication channel. A remote transponder should be within transmission range of at least two receiving sites. The system should contain two transmitters either: (1) located at one site with the capability of interrogating all transponders or (2) dispersed, with all transponders having the capability to be interrogated by two different transmitters.

The transmitter in a Type 4 system should maintain a status that permits immediate use at all times. Facilities should be provided in the central station to operate any off-line transmitter at least once every 8 hours.

Failure of any one of the receivers should not interfere with the operation of the system from the other receiver. Failure of any receiver should be annunciated. A physically separate communication channel should be provided between each transmitter/receiver site and the central station. More than one control site safeguards against damage caused by lightning and minimizes the effect of interference on the receipt of signals.

A Type 5 system has a single control site with a minimum of one receiving site and one transmitting site. The sites can be housed together.

The capacities of radio multiplex systems are based on the overall reliability of the signal receiving, processing, display, and recording equipment at the central station and the capability to transmit signals during adverse conditions in the signal transmission facilities. Allowable capacities are listed in NFPA 72, *National Fire Alarm Code*. The capacity of a system unit is unlimited when the signal receiving, processing, display, and recording equipment is duplicated at the central station and a switchover can be accomplished in a maximum of 30 seconds with no loss of signals during this period.

Local Systems

Wireless transmission is used in local protective signaling systems between the initiating devices and the control panel. To facilitate this, primary (dry cell) batteries can be the sole source of power for the initiating devices; these are usually low-power devices.

Power Supplies: A primary (dry cell) battery can be used as the sole power source of a low-power wireless initiating device transmitter, under certain conditions.

Each transmitter should serve only one device and be individually identified at the receiver/control unit. The battery should be capable of operating the wireless initiating device transmitter for at least one year before the battery depletion threshold is reached. A battery depletion signal should be transmitted before the battery is so low that it cannot support alarm transmission after seven additional days of normal operation. This depletion signal should be distinctive from all other signals; visibly identify the affected initiating device transmitter; and, if silenced, automatically re-sound at least once every four hours. A trouble signal identifying the affected initiating device transmitter should sound at its receiver/control unit when catastrophic (open or short) battery failure occurs. If silenced, the trouble signal should automatically re-sound at least once every four hours. Any failure of a primary battery in an initiating device transmitter should not affect any other initiating device transmitter.

Supervision: Low-power wireless transmitting devices should be specifically listed for a particular transmission method and highly resistant to misinterpretation of simultaneous transmissions and to interference, e.g., impulse noise and adjacent channel interference.

Transmission performance can be tested using UL 985, *Household Fire Warning System Units*, and UL 1023, *Household Burglar-Alarm System Units*.

Occurrence of any single fault that disables transmission between any wireless initiating device(s) and the receiver/control unit should cause a trouble signal within 200 seconds, except when allowed otherwise by Federal Communications Commission (FCC) regulations. A single fault on the signaling channel should not cause an alarm signal.

Normal periodic transmission from a wireless initiating device should provide additional assurance of successful alarm transmission capability by transmitting at a reduced power level or by other means.

Removal of a wireless initiating device from its installed location should cause immediate transmission of a distinctive supervisory signal indicating its removal and identifying the individual affected device.

Reception of any unwanted (interfering) transmission by a retransmission device (repeater) or by the main control unit for a continuous period of more than 20 seconds should cause an audible and visible trouble indication at the main control unit, identifying the specific trouble condition present.

Alarm Signals: When operated, each wireless initiating device transmitter should automatically transmit an alarm signal. (This is not intended to preclude the inclusion of verification and local test intervals prior to alarm transmission.) Each wireless initiating device in alarm should automatically repeat alarm transmissions at a maximum 60-second delay interval until the initiating device is returned to its normal condition. Fire alarm indicating signals should have priority over all other signals.

The system should be arranged to respond with a minimum delay to an alarm signal from a wireless initiating device transmitter. There should be a maximum 90-second allowable response delay from the first alarm transmission to receipt and display by the receiver/control unit. An alarm signal from a wireless initiating device should latch at its receiver/control unit until manually reset, and should identify the particular wireless initiating device in alarm.

Remote Station Systems and Proprietary Systems

The three retransmission methods in order of preference are: (1) a dedicated circuit independent of any switching network; (2) a one-way (outgoing only) telephone at the remote station that utilizes the commercial dial network, or other method acceptable to the local authority; and (3) a wireless retransmission in remote station systems that can transmit an alarm signal from the remote station to the fire department and consists of a private radio system that uses the fire department frequency (when permitted).

Wireless transmission can also be used in proprietary systems as the signaling channel. Where a private radio is used as the signaling channel, appropriate supervised transmitting and receiving equipment should be provided at central supervising, satellite, and repeater stations.

Certain conditions are necessary for the central supervising, satellite, and repeater station radio facilities where more than five buildings or premises are protected or where 50 initiating devices or initiating device circuits are being serviced by a private radio carrier. These include:

1. Dual-supervised transmitters installed and arranged for automatic switching from one to the other in case of trouble. Where personnel manage transmitters 24 hours a day, switchboard facilities can be manually operated if the switching can be carried out within 30 seconds. When the transmitters are located elsewhere, the circuit extending between the central supervising station and the transmitters should be a supervised circuit. Transmitters should be operated on a two-to-one time ratio basis within each 24-hour period.
2. Dual receivers installed with a means for selecting a usable output from one of the two receivers. Failure of either receiver should not interfere with the operation of the system by the other operating receiver; failure of either receiver should be annunciated.
3. Central, satellite, and repeater station radio transmitting and receiving equipment controlled and supervised for wireless transmission at the central supervising station.

Public Fire Service Communication Systems

The public fire service often uses wireless radio communication systems between coded radio fire alarm (street) boxes and the communication center. The primary use of radio communication systems is to dispatch fire companies upon receipt of an alarm at the communication center. Dispatching is done from the communication center to fire stations, mobile fire apparatus, and portable equipment. Radio devices are increasingly used to alert off-duty personnel through pagers, and those on the fireground. NFPA 1221, *Standard for the Installation, Maintenance, and Use of Public Fire Service Communication Systems*, describes in detail the requirements for each use of radio equipment in public fire service communication systems.

Coded Radio Reporting Systems

Radio Box Channel Frequency: A maximum of 500 boxes should be permitted on a single radio channel. All coded radio box systems should constantly monitor the frequency(ies) in use. Both an audible and visual indication of any sustained carrier signal (when in excess of 15 seconds) should be provided for each receiving system at the communication center.

When a box message signal to the communication center (or acknowledgment of message receipt signal from the communication center) is repeated, associated repeating facilities should meet local requirements.

Boxes: Coded radio fire alarm boxes should be designed and operated in compliance with all applicable Federal Communications Commission (FCC) rules and regulations. Boxes should provide at least three specific and individually identifiable functions to the communication center in addition to the box number: (1) test, (2) tamper, and (3) fire. Boxes should transmit to the communication center no fewer than one round for test, one round for tamper, and three rounds for fire. The FCC requires that, "Except for test purposes, each transmission must be limited to a maximum of 2 seconds and may be automatically repeated not more than two times at spaced intervals within the following 30 seconds; thereafter, the authorized cycle may not be reactivated for 1 minute."[1]

When multi-function boxes are used to transmit request(s) for emergency service or assistance to the communication center, each such additional message function should be individually identifiable. Multi-function boxes should be so designed that the loss of supplemental or concurrently actuated messages is prevented.

• *Facilities and Equipment for Receipt of Box Alarms:* Alarms from boxes should be automatically received and recorded at the communication center. Receipt of an alarm should be indicated by a permanent visual record and an audible signal. The permanent record should indicate the exact location from where the alarm was transmitted, and date and time of receipt of each alarm.

There are two basic types of coded radio reporting systems: (1) Type A and (2) Type B. For each frequency used on a Type A system, two separate receiving networks should be provided. Each should include an antenna, audible alerting device, receiver, power supply, signal processing equipment, a means of providing a permanent graphic recording of the incoming message (timed and dated), and other associated equipment. Both systems and associated equipment should be in operation simultaneously; facilities should be so arranged that a failure of either receiving network will not affect the receipt of messages from boxes. (See Figure 4-9.)

When a polling device is incorporated into the receiving network to allow remote/selective initiation of box tests, a separate such device should be included in each of the two required receiving networks. The polling devices should be configured for automatic cycle initiation in their primary operating mode, capable of continuous self-monitoring, and integrated into the network(s) to provide automatic switchover and operational continuity in the event of failure of either device.

Test signals from boxes need not include the date as part of the permanent record, provided that the date is automatically printed on the recording tape at the beginning of each day.

For each frequency used on a Type B system, a single complete receiving network should be installed in each fire station. If alarm signals are transmitted to a fire station from the communication center using the coded radio-type receiving equipment in the fire station to receive and record the alarm message, a second receiving network should be at each fire station and employ a frequency other than that used for the receipt of box messages.

The power supplied to all required circuits and devices of the system should be supervised. Receipt of alarms depends upon radio repeaters. These repeaters should be provided with dual receivers and transmitters. Failure of the primary transmitter or receiver should cause an automatic switchover to the secondary receiver and transmitter. Where repeater controls are supervised by personnel on 24-hour duty, manual switchover facilities can be provided if switchover is performed within 30 seconds.

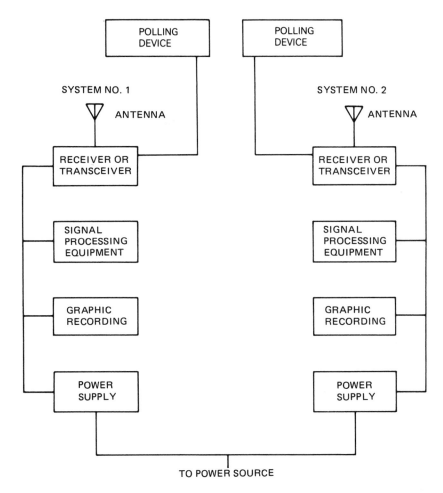

Figure 4-9. Type A coded radio reporting system. Polling is only required for transpondance (two-way) type systems.

Radio Dispatch Systems

Another use of wireless transmission for public fire service communication systems is the radio dispatch system. Two separate dispatch circuits should be provided for transmitting alarms: (1) supervised and (2) unsupervised. A circuit terminating at a telephone instrument only is not considered as either type of required dispatch circuits.

Supervised dispatch circuits should consist of one of the following:

1. A supervised wired circuit (see Figure 4-10);

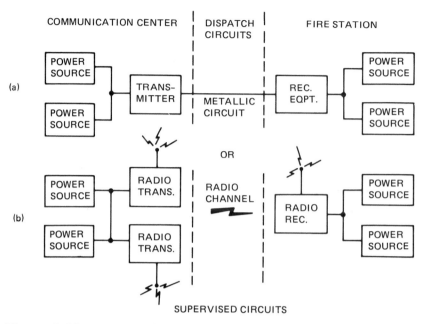

Figure 4-10. Supervised wire circuits on dispatch systems.

2. A radio channel with duplicate base transmitters, receivers, microphones, and antennae;
3. A microwave-supervised carrier channel;
4. A polling or self-interrogating radio or microwave radio system with duplicate base transmitters, and equipment necessary for redundancy; or
5. A properly arranged and supervised telephone circuit.

Unsupervised dispatch circuits can be either a wired circuit or a radio channel. (See Figure 4-11.) Unsupervised radio does not require duplicate facilities. If both supervised and unsupervised dispatch circuits are radio, separate radio frequencies should be provided. These radio dispatch channels should be separate from channels used for routine or fireground communciations. Jurisdictions that receive fewer than 600 alarms per year do not, in general, need unsupervised dispatch circuits.

Dispatching Mobile and Portable Equipment: The communication center should be equipped for radio communication with fire apparatus. All fire apparatus and other fire department emergency vehicles should be equipped with two-way radios that are FCC-type accepted or approved.

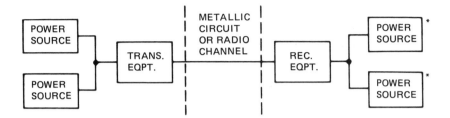

*POWER SOURCES REQUIRED IF NOT PROVIDED FROM THE
COMMUNICATION CENTER

Figure 4-11. Unsupervised circuits on a radio dispatch system.

A separate frequency should be provided for fireground communications for jurisdictions or multiple jurisdictions on the same channel that receive 2,500 or more alarms per year, or when multiple jurisdictions share a common radio frequency.

Fire portable transceivers should be capable of multiple frequency operation, enabling a fireground radio network to be organized independent of normal dispatch channels. Fire portable radios should be capable of continuous tone-coded squelch (CTCS) or continuous digital coded squelch (CDCS). When data is transmitted from fire portable transceivers, the radio should transmit data without distortion, and the equipment so designed to ensure full data stream transmission at full power.

Where fire portable transceivers are used in a fire dispatch system, they should operate properly within the dispatch area without the use of mobile RF amplifers.

Scanning devices, if used, should have an automatic priority feature whereby the radio will automatically revert to its primary channel whenever the channel is being used. Scanning devices should have a manual lock position solely to lock the receiver on its primary channel.

Miscellaneous Radio Devices: When radio home-alerting receivers, handheld units, pocket pagers, and similar radio devices are also used to receive fire alarms, or are used on the fireground, they should meet certain requirements.

Fire portable radio equipment should be manufactured for the environment in which it will be used, water-resistant, and sized for one-hand operation. Fire portable radio transceivers should not be placed into transmit mode except by the operator on a mechanically guarded switch. The operator should also be able to change channels on the multiple-frequency transceiver while wearing gloves.

The single-unit recharger for fire portable radios should be capable of recharging fully while the radio is in the receiving mode. Radio pocket pagers that are powered by replaceable batteries should indicate audibly before the battery is incapable of operating the pager.

OPTICAL FIBER TRANSMISSION

Optical fiber technology, or fiber optics, is cost effective in many different applications, including fire alarm signaling systems. Advantages of fiber optics include its high band width, low loss, small size, light weight, noise immunity, and transparency. It is also a noninterfering-type, highly secure method of transmission. On the other hand, fiber optic transmission requires active electro-optic transceivers, and is an expensive method initially. There are also nonstandard products, e.g., fiber, cable, and connectors, used in this new technology.

Principles of Fiber Optics

An optical fiber, in a sense, is a "light pipe," where particles of light are launched into the pipe at incredible speeds. (See Figure 4-12.) These particles of energy, called photons, must flow (or propagate) through an intricate system of pipes without leaking out at the interconnecting fittings (connectors) and without excessive spillage at the input and output points.

Conserving flow pressure depends on the number of photons that go completely through the pipe network without getting lost. Photons can sometimes be destroyed or absorbed by contaminates in the pipe, scattered through the walls of the pipe, or lost at the fittings.

The fiber optic link in a lightwave system consists of three basic elements: (1) transmitter, (2) cable, and (3) receiver. (See Figure 4-13.)

In fiber optics, the fiber is a waveguide. Energy in the fiber is sent through the structure by total internal reflection, as opposed to electricity which travels along the surface of the conductor. Light passing through two different but transparent media is subject to phenomena called refraction and reflection. Whenever a ray of light is incident on a boundary that separates two different media, part of the ray is reflected back into the first medium while the remainder is bent (refracted) in its path when it enters the second medium. (See Figure 4-14.) This creates a critical angle. This angle is defined as the exact

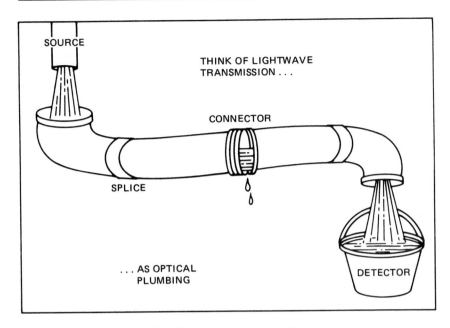

Figure 4-12. Example of lightwave transmission.

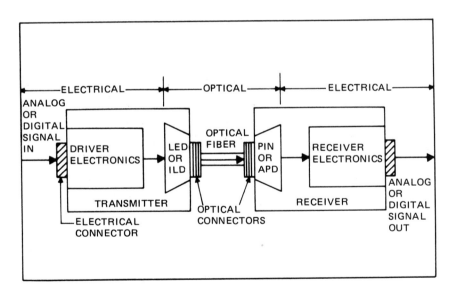

Figure 4-13. Basic lightwave system.

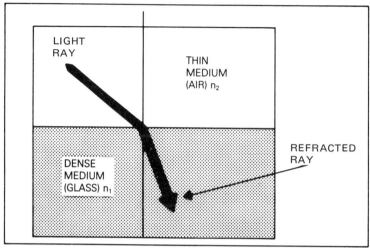

n₁ and n₂ = INDICES OF REFRACTION.

Figure 4-14.   Fiber optic light ray bending.

angle when a light ray shifts from a state where some of its energy is refracted and some reflected, to a state where all the energy is reflected. (The exactness of the critical angle is a function of the ratios of the indices of refraction of the two transparent media that the ray passes through.)

External reflection occurs when a light ray in a low-index medium, e.g., air, bounces off a high-index medium, e.g., water or mirror. (See Figure 4-15.) Internal reflection, which is more efficient, occurs when a light ray in a high-index medium, e.g., water, bounces off the interface to the low-index medium, e.g., air. (See Figure 4-16.)

A ray of light traveling in the core of the fiber is partially refracted into the clad glass unless the ray angle is less than a critical value. (See Figure 4-17.)

Fibers can be identified by the types of paths that the lightwave rays (or modes) travel within the fiber core. There are two types of fibers normally used in lightwave communications: (1) multimode and (2) single-mode. (See Figure 4-18.)

Further, there are two types of multimode fibers: (1) step index and (2) graded index. The step index multimode fiber derives its name from the sharp step-like difference in the refracted index of the core and the cladding. (See Figure 4-19.) Step index fibers are used primarily in image transport applications, such as medical and industrial remote viewing; multimode step index fibers are limited in bandwidth capabil-

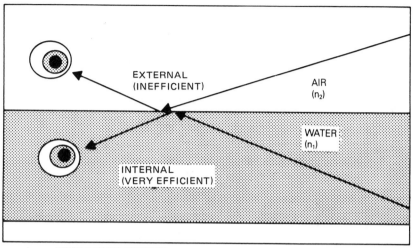

n₁ and n₂ = INDICES OF REFRACTION.

Figure 4-15. Fiber optic internal/external reflection.

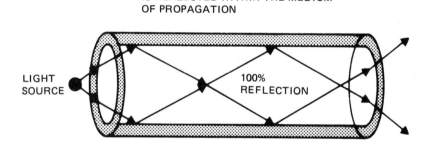

Figure 4-16. Fiber optic total internal reflection.

ities and are mostly not used today. For multimode graded index fibers, the lightwaves or rays are also guided down the fiber in multiple pathways. But unlike step index fibers, the core contains many layers of glass, each with a lower index of refraction traveling outward from the centerline of the fiber.

Grading causes the lightwaves to increase in speed as they are being refracted toward the fiber centerline and travel through the outer layers. This effectively matches the time of flight to those rays traveling the shorter pathways directly down the axis of the fiber.

PRINCIPLE OF TOTAL INTERNAL
REFLECTION FOR OPTICAL WAVEGUIDES

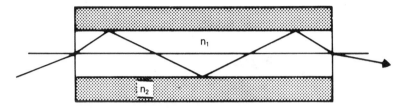

WHERE:
n = INDEX OF REFRACTION
$n_1 > n_2$ GIVES TOTAL INTERNAL REFLECTION

THIS IS A VERY EFFICIENT MECHANISM, WHEREIN:
TYPICAL MIRROR (METALIZED REFLECTION) = 95%
TOTAL INTERNAL REFLECTION GIVES 99.9999%.

Figure 4-17.   Principle of total internal reflection.

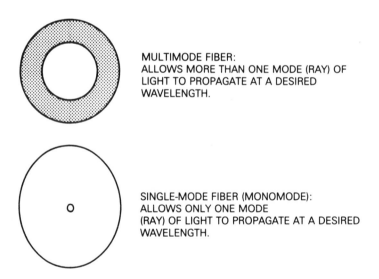

MULTIMODE FIBER:
ALLOWS MORE THAN ONE MODE (RAY) OF
LIGHT TO PROPAGATE AT A DESIRED
WAVELENGTH.

SINGLE-MODE FIBER (MONOMODE):
ALLOWS ONLY ONE MODE
(RAY) OF LIGHT TO PROPAGATE AT A DESIRED
WAVELENGTH.

Figure 4-18.   Multimode and single-mode fibers.

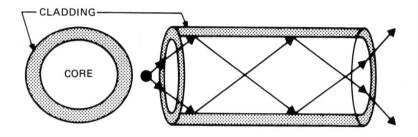

CORE: THIS IS THE CENTRAL TRANSMITTING PORTION OF
THE FIBER, EITHER GLASS OR PLASTIC. LIGHT RAYS ARE
TRAPPED IN THIS MORE OPTICALLY DENSE MEDIUM BY
TOTAL INTERNAL REFLECTION. THE LARGER THE CORE, THE
MORE LIGHT THAT WILL BE TRANSMITTED.

CLADDING: THIS OUTER LAYER HAS A LOWER REFRACTIVE
INDEX THAN THE CORE, AND CREATES THE CONTAINMENT
LAYER FOR LIGHT GUIDING IN THE FIBER.

Figure 4-19. Basic multimode step index fiber structure.

Fiber Cable

Fiber cable vastly multiplies the capacity of existing ducts. It is especially attractive where the expense of digging new conduit routes, such
as in aerospace or automotive facilities and/or existing road networks,
is prohibitive.

A single fiber can replace any one or combination of several copper
cables. The cable can be installed underground, aerially, or in ducts, and
avoids transmission problems caused by lightning, ground faults, radio/
television interference, and cable corrosion. Fiber cable does not need
special shielding conduits or grounding, because it provides extra lightning protection for computers and other equipment. Unlike copper
cables, optical fibers because of their dielectric nature do not act as
antennae or conductors if lightning strikes. Because lightwave communication is accomplished by the transmission of light energy through a
glass waveguide as opposed to the movement of electrons over a conductor, spark generation is not a problem with fiber cable.

Fiber optic cables can be run almost anywhere without concern for
communication interference or cross-talk factors. Fiber optic cable can run
directly through hazardous areas without special expensive protection.

Installation and Use: Fiber optic cables should be installed and used in
accordance with local codes. (See Article 770 of NFPA 70, *National
Electrical Code*, for detailed instructions.) The cables should be tested

for use in fire alarm signaling systems, preferably following requirements in UL 910, *Test Method for Fire and Smoke Characteristics of Electrical and Optical Fiber Cables.* Cables are normally specified to provide a stated loss and bandpass capability over a wide variety of operating conditions. Various standards organizations have developed test methods to simulate possible stresses to a cable. The Electronic Industry Association (EIA) in Alexandria, VA, has developed a comprehensive set of test methods that most manufacturers use to test their optical cables.

Meeting appropriate test requirements is not an automatic process, so the designer should work with a reliable cable supplier. The designer should understand that the cable can pass the burn test but not offer acceptable optical and mechanical performance.

The fiber optic system must be electrically compatible with the input-output characteristics with which it interfaces. Fiber optic systems are no more or less sensitive to environmental problems than ordinary copper or radio-based electronics, so it is equally important to consider the environmental conditions in which the fiber optic cable will perform.

Fiber Optic Circuits

Optical fibers are sized 62.5/125 or 100/140 for use with fiber optic communication systems. The first number refers to the diameter of the fiber in microns, while the second number is the clad diameter in microns. Each leg of a fiber optic link requires the use of two cables for duplex operation. These cables should be listed for use, such as in plenums, vertical risers, and general building installation, and must comply with any additional requirements of the authority having jurisdiction.

Fiber Optic Connectors

Fiber optic cables generally use "ST" connectors for 62.5/125 or 100/140 cable. The ST connector is a standard connector of the ferrule type, which holds and aligns the cable during installation. Some manufacturers offer fiber optic cables with factory-installed connectors. If factory-installed connectors are used, a detailed review of conduits, raceways, and cable trays must be made to determine cable length, and to verify the factory-installed connectors will not be a problem during cable installation.

During field installation, a prime concern is the prevention of stress during connector installation. If the connectors are under stress condition during installation, the stress can be transmitted to the cable, damaging the polished condition of the fibers, thus creating an intermittent problem condition. The manufacturer's installation procedures must be followed precisely for trouble-free results.

Another concern during field installation of fiber optic cables is the bend radius. The minimum bending radius varies with the manufacturer and diameter of the optical fiber. The minimum bend radius will increase as the length of the cable pull is increased. This is to reduce the friction associated with the bends and the distance, and to limit the tension stress applied to the cable. As a general rule, a minimum bend radius of 1.5 inches (38 mm) for both 62.5/125 and 100/140 cables can be used as a guide.

Two other stress conditions, i.e., tension load stress and static stress, must be considered during installation to minimize problem conditions.

Fiber Optic Attenuation

Each fiber link module utilizing fiber optic cable has a "fiber optic attenuation" or "fiber optic budget" that indicates the maximum light attenuation permitted in the fiber cable that will continue to sustain reliable communication. In order to verify that the maximum level has not been exceeded, calculations should be prepared for each fiber link. The total link loss should not exceed the fiber optic budget for the module.

Total fiber optic link attenuation is determined based on the following individual factors, wherein the total must be less than the fiber optic budget for the system to operate properly:

1. The fiber optic cable loss per unit distance,
2. The length of the fiber optic cable,
3. The number of splices in the link; as a general rule, each splice will have a 2 dBA attenuation factor, and
4. An estimate of the number of future splices in the link.

SIGNAL TRANSMISSION COMPATIBILITY

Compatibility—the ability of equipment produced by one manufacturer to operate properly with that of another manufacturer—is critical to signal transmission in a signaling system. If the control panel is manufactured by one company, compatibility must be determined with other manufacturers' detectors, indicating appliances, annunciator panels, etc. The key to system compatibility is the installation wiring diagram furnished with the equipment. If a control panel alarm initiating circuit is compatible only with certain devices, those devices and pertinent installation instructions should be specified on the diagram. A circuit extending from a control panel that contains only an electrical rating and does not specify specific devices will be compatible with any device that has a similar electrical rating.

For example, when a 4-wire smoke detector senses smoke, the smoke detector acts as a switch across the alarm initiating circuit while two other wires provide power to the smoke detector. The control panel initiating circuit does not differentiate between the signal from a smoke detector switch closure (relay contact) and a signal from a heat detector or a manual fire alarm box. Since a 4-wire smoke detector does not impose any load on the alarm initiating circuit of the control panel, it can be used with an electrically compatible fire alarm system.

In contrast, a 2-wire smoke detector is powered by and sends signals over the same two wires of an alarm initiating circuit. Compatibility between 2-wire smoke detectors that receive power from the initiating device circuit of a fire alarm system control panel and the panel depends upon the interaction between the circuit parameters, e.g., voltage, current, frequency and impedance of the detector, and the initiating device circuit. A 2-wire smoke detector must be compatible, therefore, not only with its control panel, but also with other 2-wire detectors on the circuit.

Because a 2-wire detector obtains power from the initiating device circuit of a system control panel, its operation depends on the characteristics of the circuit to which it is connected, as the detector imposes a resistive and capacitive load on the circuit.

Compatibility between the detector and the control panel will ensure that sufficient power is furnished to the detector at startup, in the standby (supervisory) mode, and that the change in status of a smoke detector when it goes into alarm will be sensed at the control panel. If the smoke detector has an integral audible signal, annunciator light, or supplementary relay, the control panel must also be able to furnish the power for these functions.

To be listed, smoke detectors and a compatible control panel are tested for all of these conditions to ensure proper signal transmission when the system is placed into service.

SIGNAL VERIFICATION VS. FALSE ALARMS

When a transmission circuit is established, it is important to verify that a signal is truly an alarm and not a false alarm. Some methods used to reduce the number of false alarms from smoke detectors in equipment and in the field are as follows.

False Alarm Reduction in Equipment

1. Several alarm pulses can be counted before an alarm is signaled on smoke detectors with pulsing normal operation.
2. Detectors that have either an analog output to the control panel or a field sensitivity adjustment on the detector can have the sensitivity (alarm set point) decreased within limits.
3. A delay-reset time of 5 to 30 seconds can be provided where the signal is held and the detector then reset. If the detector still senses smoke for a period of time after the delay-reset, the alarm signal is transmitted.
4. On a voluntary basis, Underwriters Laboratories Inc. permits reduction of the sensitivity of all detectors at the factory by 50 percent or 1 percent per ft (3.3 percent/m) obscuration, whichever is less.

False Alarm Reduction in the Field

1. A thorough evaluation of the building should be performed to ensure that the smoke detectors are placed away from areas that would normally be humid, dusty, smoky, or insect-laden. Detectors should also not be located in areas with temperature extremes, excessive RF noise, or other electrical noise on the power line or in the area. Any of these could be the source of false alarms.
2. Smoke detectors should be installed in proper relation to all supply and return air ducts.
3. Installation of the smoke detector heads can be delayed until all of the building construction and plaster dust has been removed.
4. A regular cleaning of all smoke detectors should be performed as recommended by the manufacturer.
5. The detector sensitivity should be regularly checked and recorded. Detectors that have drifted away from their normal setting should be replaced.

6. Smoke detectors can be installed away from smokers, if possible, or smoking prohibited in the area.

In addition to the above methods of reducing false alarms, several industry programs will help to reduce false alarms. The National Institute for Certification in Engineering Technologies (NICET) in Alexandria, VA, has established a national certification program for technicians in the fire alarm systems field. The program covers types of systems, job specifications, codes and standards, surveys, layout drawings, installation, wiring, system operation, and testing and maintenance of fire alarm systems.

Underwriters Laboratories Inc. has established a certificate service program for fire protective signaling systems. This program qualifies alarm companies to certify that a fire alarm system has been installed and tested in compliance with NFPA 72, *National Fire Alarm Code.*

BIBLIOGRAPHY

NFPA Codes, Standards, Recommended Practices, and Manuals. (See the latest *NFPA Catalog* for availability of current editions of the following documents.)

NFPA 70, *National Electrical Code.*

NFPA 72, *National Fire Alarm Code.*

NFPA 1221, *Standard for the Installation, Maintenance, and Use of Public Fire Service Communication Systems.*

Reference Cited

1. FCC *Rules and Regulations*, Vol. V, Part 90, March 1979, Washington, DC.

Additional Reading

UL Publications. Underwriters Laboratories Inc., 333 Pfingsten Rd., Northbrook, IL, 60062.

UL 910, *Test Method for Fire and Smoke Characteristics of Electrical and Optical Fiber Cables.*

UL 985, *Household Fire Warning System Units.*

UL 1023, *Household Burglar-Alarm System Units.*

5

Signal Processing

INTRODUCTION

This chapter describes automatic and manual processing of fire alarm signals for the various types of protective signaling systems. Combined manual and automatic processing such as that found in guard's tour supervisory service; sprinkler system waterflow alarm and supervisory signal service; and trouble signals are also included. See also Chapter 6, "Electrical Supervision," and Chapter 7, "Signal Notification," for further information.

SIGNAL PROCESSING

Processing is the dynamic link between signal input and system response. Signal processing is an interpretive function, governing the response of a unit or an entire fire alarm signaling system to the various inputs received by that unit or system. A control panel will typically receive numerous, varied signal inputs; outputs can vary as widely, depending on the application of the alarm system.

In a simplified application, such as for a local protective or household warning system, the control panel may dictate energization of general evacuation output signals. With a more complex system, e.g., proprietary, remote station, or central station system, several response (output) modes may need to be activated for certain types of inputs. The interpretive processing routines to match appropriate outputs to inputs can be established through fixed electrical or electronic circuitry, or controlled with flexible, complex computer programs.

TYPES OF SIGNALS PROCESSED

Manual Fire Alarm Signals

Requirements for manual fire alarm signals are addressed in NFPA 72, *National Fire Alarm Code*, and in Chapter 7 of NFPA *101*®, *Life Safety*

Code®. Manual fire alarm signals in a protective system are typically initiated from manual fire alarm boxes located throughout the protected facility. Signals can also be initiated outside the protective system, e.g., from a municipal fire alarm box. Processed outputs for manual signals vary according to the type of protective system.

Evacuation signals, whether audible (tonal or voice generated) or visible, are signal outputs in the local protective and emergency voice/alarm systems. Various prerecorded tone and voice messages can also be considered as emergency system outputs.

Manual initiation of an alarm signal in an auxiliary, remote station, and public fire service system results in the transmission of signals to a remote point for subsequent fire department interpretation and action. Signals can be transmitted indirectly, or directly to the fire department from an on-premise alarm system.

Manual signals are processed in a similar manner for central station and proprietary systems, except that the signal is processed through private system equipment with possible retransmission to a public system. Unless system design dictates, there is no distinction made between manual and automatic fire alarm signals.

Some of the differences among manual system responses are:

1. An auxiliary system does not normally use a coded alarm signal. The rationale for this is that auxiliary systems have historically been simple noncoded latching circuits that did not transmit signals in a serial method to a public fire department.
2. Manual signal initiation is not part of household fire warning equipment. Since most fires occur while the occupants of the household are sleeping, initiation and detection are the only necessary functions of household fire warning equipment.
3. NFPA 72, *National Fire Alarm Code*, and NFPA 1221, *Standard for the Installation, Maintenance, and Use of Public Fire Service Communication Systems*, assume that other system units or interfaces have processed manual signals prior to retransmission. Noncoded manual alarm inputs are processed in a manner similar to that for automatic alarm inputs.

Automatic Fire Alarm Signals

Automatic fire alarm signals for each system are addressed in NFPA 72, *National Fire Alarm Code*. Automatic alarms are typically initiated by smoke and heat detectors, both of which can be installed on separate circuits or integrated into the same circuit. Smoke and heat detectors can be installed and maintained in protective signaling systems, according to Chapter 5 of NFPA 72, *National Fire Alarm Code*, with testing requirements addressed in Chapter 7. NFPA 72, *National Fire Alarm Code*, refers to special connections for use of integral trouble contacts located on the automatic signal initiating devices. These normally closed contacts should be connected electrically at the end of the initiating circuit to prevent disconnection of initiating devices.

The differences in automatic alarm signal generation and processing depend on the type of protective signaling system used. An evacuation alarm signal is required for building occupants when a local protective system is specified. Incoming automatic signals for central station and proprietary signaling systems are required to be identified and processed by the protected zone or other system subdivision within the protected building.

Automatic devices connected to initiating circuits are required by Chapter 3 of NFPA 72, *National Fire Alarm Code*, to be latching, but nonlatching circuits are permitted for smoke detectors. Nonlatching smoke detector circuits are allowed by NFPA 72 because of the transient nature of residential conditions, e.g., cooking, humidity or condensation, or other factors, that can easily activate a smoke detector.

Primary power, when restored, must also restore systems to their normal supervisory condition. Central station and proprietary systems have a large amount of information contained within a common system. A rapid, more discriminating means to identify changes in system status is required in central station systems, since identifying and interpreting signals is important to public fire department response.

TYPES OF PROCESSING

Guard's Tour Supervisory Service

In guard's tour service, a supervisory service signal is typically initiated through mechanical key-operated tour stations. These signals can also be generated electronically or optically for encoding and decoding. "Encoding" is defined as processing of a signal such as that initiated by a switch contact and subsequent transfer of the signal into information that can be interpreted by a machine. "Decoding" is the translation of this information back into signals understandable by human beings.

Guard's tour supervisory signals are not transmitted to public fire departments, since the signals are additional communications for public fire service receiving stations.

Guard's tour supervisory signaling equipment generally tracks all stations within the tour schedule; all stations must report in this type of service. The beginning and end of a tour schedule must register at some remote point, and intermediate nonrecording signals must operate in fixed succession. A guard's tour system can have an exception reporting package in which all stations must report; such a system also has a delinquency signal to identify any station that has not been operated.

Sprinkler System Waterflow Alarm and Supervisory Signal Service

In general, a sprinkler system waterflow alarm and supervisory signal service in any protective signaling system must be able to interpret the type of system signal and the particular element that has operated. Additionally, the supervisory service must identify restoration of the element to its normal position.

Waterflow alarm supervisory signals consist of one or more of the following: gate valve supervision, pressure supervision, water level supervision, water temperature supervision, and pump supervision. The waterflow alarm signals are the only supervisory signals that can be transmitted and processed by auxiliary systems. If additional supervisory services are desired, another type of protective system should be specified.

Trouble Signals

To varying degrees, all systems interpret factors, i.e., "trouble signals," that inhibit alarm transmission or annunciation. Many functions and/or circuits are supervised against loss of capability, depending on the nature of the system and the complexity of the information processed by that system. In some cases, automatic or manual control of various system functions permits emergency operation under faults that could debilitate the entire system.

Protective signaling systems share certain characteristics concerning trouble signals. All systems interpret loss of primary (main) power as a trouble condition, which is usually indicated by an audible signal. Loss of initiating circuits is also a trouble condition, and protective systems process this signal information accordingly. Up to five styles of fault recognition can be specified for a proprietary protective signaling system.

Loss of indicating or evacuation circuits is a trouble condition in local protective and emergency voice/alarm communication systems. When central station, remote station, or proprietary systems have an evacuation alarm signal, these integrated circuit systems should comply with NFPA 72, *National Fire Alarm Code*, concerning trouble signals for loss of both evacuation and indicating circuits.

Trouble signals for signaling line circuits are covered in Chapter 4 of NFPA 72, *National Fire Alarm Code*. This standard requires that the central station, remote station, and proprietary systems recognize a single open and single ground fault as an abnormal (trouble) condition, but not as an alarm or fire.

The most comprehensive process recognition design is found in the proprietary systems, in which prescribed responses are defined for single and multiple faults on both the initiating device and signaling line circuits. Tables 6-1 and 6-2, in Chapter 6, show the performance and capacities of these circuits. Styles A through C in Table 6-1 are considered Class B circuits, and Styles D and E as Class A circuits. The table breaks out each style by specific performance, since the circuits differ slightly in individual performance.

A central station system is also assigned system capability based on various responses, but to a lesser degree than for proprietary systems. NFPA 72, *National Fire Alarm Code*, specifies appropriate responses for virtually any kind of initiating device circuit or signaling line circuit. While the various circuit styles do not necessarily represent an ascending order of complexity, the system processing/response implies that the higher lettered styles can accommodate greater distribution and concentration of connected devices. There is also an ascending order of complexity and capability for system processing/response in a central station system.

BIBLIOGRAPHY

NFPA Codes, Standards, Recommended Practices, and Manuals. (See the latest *NFPA Catalog* for availability of current editions of the following documents.)

NFPA 70, *National Electrical Code*.

NFPA 72, *National Fire Alarm Code*.

NFPA *101, Life Safety Code*.

NFPA 1221, *Standard for the Installation, Maintenance, and Use of Public Fire Service Communication Systems*.

6

Electrical Supervision

INTRODUCTION

Electrical supervision of a fire alarm signaling system is critical to smooth, uninterrupted system operation. Power supplies, system circuits, and equipment can be supervised so that an open or ground fault condition will activate a trouble signal. This chapter describes the components of a system that should be supervised, the individual circuit and system approaches to supervision, and the relation of device circuits to the fire alarm system. Information on trouble signals can also be found in Chapter 8, "Power Supplies." Additional information on supervision can be found in Chapter 5, "Signal Processing," and Chapter 7, "Signal Notification."

SUPERVISION OF POWER SUPPLIES, SYSTEM CIRCUITS, AND EQUIPMENT

Electrical supervision in fire detection and alarm systems monitors the circuit integrity of interconnecting conductors so when a single open or a single ground condition occurs that would prevent normal operation, the condition is automatically transmitted and indicated in the appropriate location. Electrical supervision does not normally include monitoring electrical conductors for a short-circuit fault or multiple ground condition. Short-circuit faults and multiple ground conditions will typically generate an alarm signal that will occur on an initiating device circuit or signaling line circuit. Use of double-loop circuits or multiple-path conductors instead of electrical supervision is not permitted by NFPA 72, *National Fire Alarm Code*.

Fire alarm systems provide several distinct types of audible signals. To eliminate confusion between "supervised" and "supervisory," NFPA 72, *National Fire Alarm Code*, uses the term "monitoring integrity of installation conductors." A "trouble signal" is required to sound upon the occurrence of a single open or single ground condition that would prevent normal operation of the system. Further, a "supervisory signal" indicates the off-normal position, or condition, of some part of a sprinkler or other extinguishing system.

All fire alarm and process monitoring alarm supervisory systems should be electrically supervised so the occurrence of a single open or a single ground fault condition in the installation wiring that prevents the required normal operation of the system or causes failure of the primary (main) power supply source is indicated by a distinctive trouble signal.

Power supplies, system circuits, and certain equipment in a fire alarm system should be supervised. Fault conditions, on supplementary circuits, need not be indicated by a distinctive trouble signal. However, a fault on these circuits should not affect system operation.

Power Supply Supervision

All sources of energy, i.e., power supply, in a fire alarm system should be supervised, except for a power supply that:

1. Operates trouble signal circuits and appliances.
2. Is used as an auxiliary means to maintain normal system operation following a trouble signal indicating when the primary (main) supply source is interrupted.
3. Supplies supplementary devices and appliances. Failure of this supplementary power supply should not prevent system operation and alarm signal initiation.
4. Is a battery lead for a float or trickle-charged battery.

System Circuit Supervision

Signal Initiating Circuits: Installation wiring for all appliances or devices that initiate or transmit signals, either manually or automatically, should be supervised. Typical appliances and devices are fire alarm boxes, fire detectors, or automatically operated transmitters. A noninterfering shunt circuit does not need to be supervised, provided that a fault condition of the shunt wiring does not affect essential operating features of a system. Supervision is also not essential for pneumatic continuous line-type rate-of-rise systems where the device wiring terminals are connected in parallel across electrically supervised circuits.

Audible and Visible Alarm Indicating Circuits: All installation wiring for operating audible and visible alarm indicating circuits should be electrically supervised. A short or fault in the audible alarm circuit should be indicated by a trouble signal. Supervision is not essential for the installation wiring of the following:

1. Audible alarm indicating appliance that is installed in the same room as a system control unit, provided the circuit conductors are installed in conduit or have equivalent protection against mechanical injury or tampering.
2. Supplementary and alarm indicating appliance, provided that a fault condition does not prevent required system operation.
3. Supplementary signal annunciator or other supplemental visible alarm appliance, provided that a fault condition does not prevent required system operation.

Equipment Supervision

Speaker amplifier and tone-generating equipment should be supervised where speakers are used to produce audible fire alarm signals. An audible trouble signal should result upon failure of the audio amplifier and/or tone-generating equipment. Tone-generating equipment and amplifying equipment need not be supervised if they are enclosed as integral parts and serve only one listed audible signaling appliance.

Installation wiring connections to alarm indicating appliances, initiating devices, and supervisory initiating devices should be supervised. Supplementary circuits that operate fan motor stops or similar equipment do not need to be electrically supervised, provided that a fault condition does not affect required signaling system operation. The trouble signal circuit does not need to be supervised.

Circuit Performance: Various initiating device circuits monitor the integrity of installation conductors by indicating a fault or trouble condition that would prevent normal operations. The capability of a system, as related to the performance requirements of these circuits, can be determined by placing circuits into "style" categories, i.e., A, B, C, D, or E.* Styles A through C identify a fault condition that could possibly affect system operation, and Styles D and E identify faulty circuits that could be conditioned either manually or automatically to operate despite the trouble or fault condition. Classifying circuits by styles allows a designer to know circuit performance before system specifications are written.

Performance of Initiating Device Circuits and Signaling Line Circuits:
Table 6-1 identifies the style categories of initiating device circuits based on their ability to indicate alarm and trouble at the central supervising station during specified abnormal conditions. Table 6-2 can be used for a similar purpose with signaling line circuits.

Users, designers, manufacturers, and the authority having jurisdiction can use Tables 6-1 and 6-2 to identify minimum performance of present and future systems by determining the trouble and alarm signals received at the central supervising station for the specified abnormal conditions.

To use Tables 6-1 and 6-2, first determine whether the initiating devices are (1) directly connected to the initiating device circuit or (2) directly connected to an initiating device circuit that in turn is connected to a circuit interface on the signaling line circuit. Next, determine the style of signaling performance required. The rows marked A through E in Table 6-1 and 0.5 through 7 in Table 6-2 are arranged in ascending order of performance and capacities.

Once the style of the system is determined, the charts (singularly or together) will specify the maximum number of devices, equipment, premises, and buildings that can be incorporated into an actual installation for a proprietary protective signaling system. Or, working in the opposite manner, a system style can be determined by knowing the number of devices, equipment, premises, and buildings in addition to signaling ability in an installation.

It is not intended that the styles be construed as "grades." One style system is not necessarily better than another; in fact, a particular style may provide more adequate and reliable signaling for an installation than a complex style. The quantities tabulated under each style do, unfortunately, tend to imply that one style is better than the one to its left. The increased quantities for the higher style numbers are based on

* *Styles A through C were previously known as "Class B" circuits, and Styles D and E as "Class A" circuits. This text will use the term "styles" as opposed to "class."

the ability to signal an alarm during an abnormal condition in addition to signaling the same abnormal condition. When the capacities are at the maximum allowed, the overall system reliability is considered to be equal from style to style.

Upon determining the style of the system, Tables 6-1 and 6-2 indicate the maximum number of devices, equipment, protected buildings, etc., that can be incorporated into an actual installation for a proprietary protective signaling system.

The number of automatic fire detectors connected to an initiating device circuit should be limited by good engineering practice. If a large number of detectors are connected to one initiating device circuit covering a widespread area, pinpointing the source of alarm becomes difficult and time consuming. On certain types of detectors, a trouble signal results from faults in the detector. When this occurs with a large number of detectors on an initiating device circuit, locating the faulty detector also becomes difficult and time consuming.

Initiating devices listed in Section E of Table 6-1 should not be combined on the same initiating device circuit except that manual and automatic fire alarm devices may be combined in the same initiating device circuit or where only one fire alarm box is required. The box may be connected to the waterflow initiating device circuit. The loading of initiating device circuits should not exceed the limits in Table 6-1.

Where occupancy conditions permit, the authority having jurisdiction may allow connection of a single intermediate fire alarm or fire supervisory control unit to a proprietary protective signaling system initiating device circuit.

A redundant signaling line circuit, leg facility, or trunk facility must meet the performance requirements of Table 6-2.

Tables 6-1 and 6-2 can be used for planning a local protective signaling system. In all cases, however, the system must be limited to one building.

Alarm indicating appliance circuit performance and capabilities are listed in Table 6-3.

Monitored Circuits: Each protective signaling system has a number of circuits connected to the system control unit. Each circuit fulfills a general system function and can be arranged in different ways. Major circuits found in protective signaling systems are basic initiating circuits, indicating appliance circuits, combination initiating and indicating circuits, and signaling line circuits.

TABLE 6-1. Performance and Capacities of Initiating Device Circuits

G = Systems with ground detection must indicate system trouble with a single ground.
X = Indication at central supervising station.
* = The number of automatic fire detectors connected to an initiating device circuit should be limited by good engineering practice. If a large number of detectors are connected to one initiating device circuit covering a widespread area, pinpointing alarm source becomes difficult and time consuming. On certain types of detectors, a trouble signal results from faults in the detector. When this occurs with a large number of detectors on an initiating device circuit, locating the faulty detector also becomes difficult and time consuming.
IDC = Initiating device circuit.
SLC = Signaling line circuit.

In the table below, ARC = "Alarm Receipt Capability During Abnormal Condition".

Style	A Alarm	A Trouble	A ARC	B Alarm	B Trouble	B ARC	C Alarm	C Trouble	C ARC	D Alarm	D Trouble	D ARC	E Alarm	E Trouble	E ARC
Abnormal Condition															
A. Single Open		X			X			X		X	X		X	X	
B. Single Ground		X			G	X		G	X		G	X		G	X
C. Wire-to-Wire Short	X			X			X	X			X		X		
D. Loss of Carrier (If Used)									X						X
E. Maximum Quantity per Initiating Device Circuit															
1. Fire Alarm															
(a) Manual Fire Alarm Boxes		2			5			5			25			25	
(b) Waterflow Alarm Devices		1			2			2			5			5	
(c) Discharge Alarm from Other Fire Suppression Systems		1			2			2			5			5	
(d) Automatic Fire Detectors		*			*			*			*			*	
2. Fire Supervisory															
(a) Sprinkler Supervisory Devices		2			4			4			20			20	

TABLE 6-1: Performance and Capacities of Initiating Device Circuits (continued)

Style	A	B	C	D	E
(b) Other Fire Suppression Supervisory Devices	2	4	4	20	20
3. Guard's Tour	1	1	1	1	1
4. Process, Security, and Other Devices in Combination with 1, 2, and 3 Above.	0	0	0	0	0
5. Process, Security, and Other Devices Not Combined with 1, 2, and 3 Above.	5	10	10	20	20
6. Buildings	1	1	1	1	1
7. Intermediate Fire Alarm or Fire Supervisory Control Unit.	1	1	1	1	1
F. Maximum Quantity of Initiating Device Circuits per Circuit Interface Between IDC & SLC					
1. Per Limits of E Above	10	10	10	10	10
2. With Following Limitations Fulfilled	10	20	20	50	50
(a) One Waterflow per IDC					
(b) Maximum of 4 Sprinkler Supervisory Devices on an IDC					
(c) Maximum of 5 Process, Security, and Other Devices on a Separate IDC					
(d) Maximum of 1 Intermediate Fire Alarm or Fire Supervisory Control Unit per IDC.					

TABLE 6-2. Performance and Capacities of Signaling Line Circuits

Legend:

- G = Systems with ground detection must indicate system trouble with a single ground.
- M = May be capable of alarm receipt with a wire-to-wire short.
- X = Indication at central supervising station.
- CSS = Central supervising station.
- IDC = Initiating device circuit.
- SLC = Signaling line circuit.

ARC = Alarm Receipt Capability During Abnormal Condition

Abnormal Condition	0.5 Alarm (1)	0.5 Trouble (2)	0.5 ARC (3)	1 Alarm (4)	1 Trouble (5)	1 ARC (6)	2 Alarm (7)	2 Trouble (8)	2 ARC (9)	3 Alarm (10)	3 Trouble (11)	3 ARC (12)	3.5 Alarm (13)	3.5 Trouble (14)	3.5 ARC (15)
A. Single Open		X			X		X	X			X			X	
B. Single Ground		X		G	X		G	X		G	X		G	X	X
C. Wire-to-Wire Short									M		X			X	
D. Wire-to-Wire Short & Open									M		X			X	
E. Wire-to-Wire Short & Ground								G	M		X			X	
F. Open and Ground							X	X			X			X	
G. Loss of Carrier (If Used)														X	

	0.5	1	2	3	3.5
H. Maximum Quantity per Signaling Line Circuit					
1. Initiating Devices (All Types)	250	250	250	300	300
2. Buildings	25	25	25	50	50
I. Maximum Quantity per Central Supervising Station (CSS)					
1. Initiating Device Circuits	500	500	500	1000	1000
2. IDC with Redundant CSS Control Equipment*	1000	1000	1000	2000	2000
3. Buildings	25	25	25	25	25
4. Buildings with Redundant CSS Control Equipment*	25	25	25	50	50
J. Maximum Quantity Using a Satellite Station					
1. Buildings per Signaling Line Circuit	25	25	25	25	25
2. Buildings per Signaling Line Circuit with Redundant Signaling Line	25	25	25	25	25

TABLE 6-2. Performance and Capacities of Signaling Line Circuits (continued)

Definitions:

- G = Systems with ground detection must indicate system trouble with a single ground.
- M = May be capable of alarm receipt with a wire-to-wire short.
- X = Indication at central supervising station.
- CSS = Central supervising station.
- IDC = Initiating device circuit.
- SLC = Signaling line circuit.

Style	4			4.5			5			6			7		
	Alarm	Trouble	Alarm Receipt Capability During Abnormal Condition	Alarm	Trouble	Alarm Receipt Capability During Abnormal Condition	Alarm	Trouble	Alarm Receipt Capability During Abnormal Condition	Alarm	Trouble	Alarm Receipt Capability During Abnormal Condition	Alarm	Trouble	Alarm Receipt Capability During Abnormal Condition
Abnormal Condition	16	17	18	19	20	21	22	23	24	25	26	27	28	29	30
A. Single Open		X			X	X		X	X		X	X		X	X
B. Single Ground		G	X		X			G	X		G	X		G	X
C. Wire-to-Wire Short		X			X			X			X			X	X
D. Wire-to-Wire Short & Open		X			X			X			X			X	
E. Wire-to-Wire Short & Ground		X			X			X			X			X	
F. Open and Ground		X			X			X			X	X		X	X
G. Loss of Carrier (If Used)		X			X			X			X			X	
H. Maximum Quantity per Signaling Line Circuit — 1. Initiating Devices (All Types)		500			500			750			1000			1000	
2. Buildings		75			75			75			100			100	
I. Maximum Quantity per Central Supervising Station (CSS) — 1. Initiating Device Circuits		1000			1000			1500			2000			2000	
2. IDC with Redundant CSS Control Equipment*		2000			2000			3000			Unlimited			Unlimited	
3. Buildings		50			50			75			400			400	
4. Buildings with Redundant CSS Control Equipment*		100			100			150			Unlimited			Unlimited	
J. Maximum Quantity Using a Satellite Station — 1. Buildings per Signaling Line Circuit		50			50			75			75			100	
2. Buildings per Signaling Line Circuit with Redundant Signaling Line		100			100			150			500			500	

*When the central supervising station multiplex control unit is duplicated and a switchover can be accomplished in not more than 90 seconds without loss of signal during this period, the capacity of the system is unlimited.

TABLE 6-3. Performance and Capabilities of Alarm Indicating
Appliance Circuits

Style	M		N		O		P	
G = Systems with ground detection must indicate system trouble with a single ground. X = Indication at protected premises. F = Operates only up to location of fault.	Trouble Indication at Protected Premises	Alarm Capability During Abnormal Condition	Trouble Indication at Protected Premises	Alarm Capability During Abnormal Condition	Trouble Indication at Protected Premises	Alarm Capability During Abnormal Condition	Trouble Indication at Protected Premises	Alarm Capability During Abnormal Condition
Single Open	X	F	X	X	X	F	X	X
Single Ground	X		X		G	X	G	X
Wire-to-Wire Short	X		X		X		X	

• *Basic Initiating Circuits:* The basic alarm initiating circuit (typically Style A, B, or C) consists of a 2-wire circuit with an end-of-line resistor. A separate circuit in the control panel monitors the field wiring between the panel and the initiating devices. Initiating devices have normally open contacts and are connected in parallel. (See Figure 6-1.) As shown, a small supervisory current normally flows through both relays (or an equivalent solid-state circuit), the field installation wires, and the end-of-line resistor. The current is sufficient to energize T_1 but not A_1. Operation of an initiating device (detector) shunts the resistor, increasing the current in the circuit, which energizes A_1 to initiate the alarm. A short between the wires or a ground fault on both wires will also cause an alarm.

A single open on the circuit would deenergize T_1 and initiate a trouble signal; it would also render inoperable all initiating devices that are electrically beyond the open as shown in Figure 6-2. When an open fault occurs, audible and visible trouble indications should operate at the control unit.

When an automatic fire detector has integral trouble contacts, contacts must be wired between the last detector and the end-of-line resis-

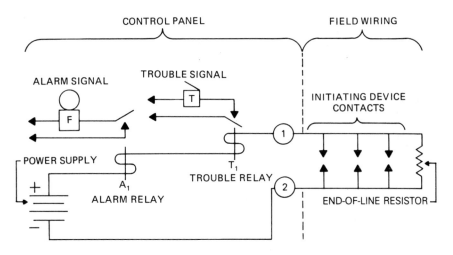

F = ALARM INDICATING APPLIANCE

Figure 6-1. Typical Style A, B, or C initiating device circuit.

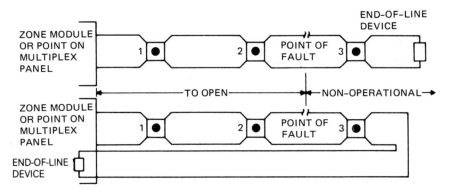

Figure 6-2. Typical Style A, B, or C initiating device circuit with open fault.

tor. This prevents the opening of a trouble contact from impairing the alarm operation of other fire detectors. (See Figure 6-3.)

To operate the system with an open circuit fault (Style D or E) on the initiating circuit, wiring would be typically arranged as shown in Figure 6-4. The circuit would be connected to a control panel designed to receive the signal despite an open on the field wiring circuit. Since the end-of-line device is at the fire alarm control panel in the form of relay T_1, it deenergizes when an open on the wire occurs, sounding the trou-

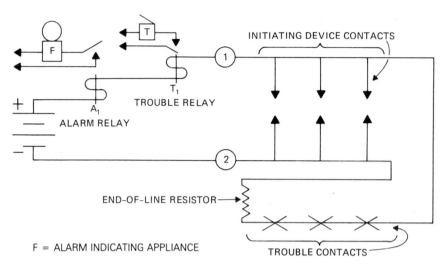

Figure 6-3. Typical Style A, B, or C initiating device circuit with trouble contacts.

ble signal "T." The panel then would be conditioned manually or automatically by the operation of S_1 and S_2 to connect terminals 1 and 4, and 2 and 3 together. The system could then operate for an alarm despite an open circuit on either or both of the initiating circuit wires. Should two opens occur on a single wire, however, such as at point "A" and "B," the device to the right of point "A" would not be able to operate the system.

Alarm initiating circuits should be wired as shown in Figure 6-5 to monitor all interconnecting wires. If it is incorrectly wired (Figure 6-6), the fire alarm system would set up as if it were in the normal supervisory (monitored) condition—but an open occurring in riser No. 1 or 2 would not be indicated at the control unit because the supervisory current would still flow through riser No. 3 and the end-of-line device. A more detailed discussion of wiring methods for fire alarm systems is found in Chapter 4.

• *Indicating Appliance Circuits:* Indicating appliance circuits can also be basic supervised circuits, or capable of being conditioned to operate despite an open circuit on the installation wiring.

A typical basic indicating appliance circuit is shown in Figure 6-7. In the normal supervisory condition, current would flow from the negative side of the power supply through the normally closed contact (A_{1b}) of the alarm relay A_1 then exits terminal 2 and flows through the end-

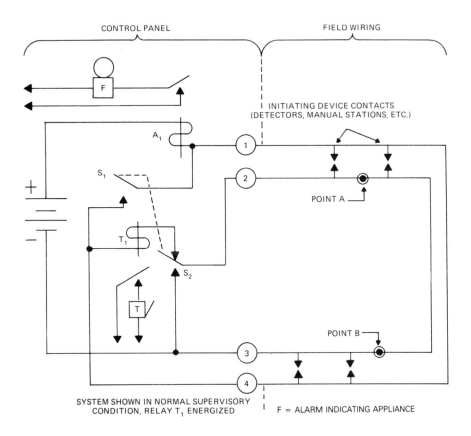

Figure 6-4. Typical Style D or E initiating device circuit.

of-line diode D_1, back through terminal 1, through the normally closed contact (A_{1a}), and through the supervisory relay coil T_1 to the positive side of the power supply.

In this supervisory condition, the D_2 diodes block the flow of current through the alarm appliances "F," preventing their operation. Sufficient current flows through relay T_1 to energize it as shown. Should an open occur in the wiring to the alarm appliances, or D_1, relay T_1 will deenergize, activating the trouble signal "T."

When an alarm occurs and alarm relay A_1 is energized, contacts A_{1a} and A_{1b} transfer from the upper, normally closed position, to the lower, normally open position. This changes the potential at terminal 1 from plus (+) to minus (-), and the opposite at terminal 2. Diode D_1 then blocks the flow of current through the end-of-line device, and the D_2

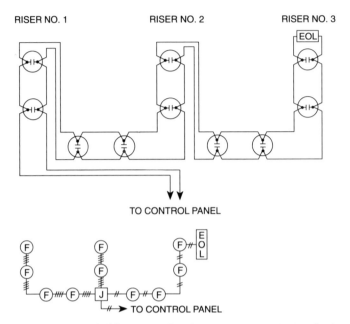

Figure 6-5. Correct field wiring of Style A, B, or C initiating device circuit.

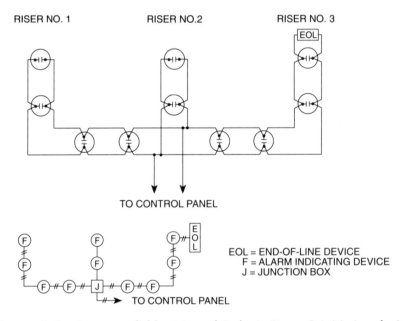

Figure 6-6. Incorrect field wiring of Style A, B, or C initiating device circuit.

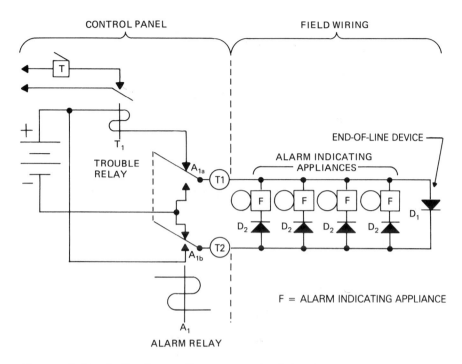

CONTROL PANEL   FIELD WIRING

Figure 6-7. Indicating appliance supervised circuit.

diodes conduct current to "F," causing the indicating appliances to operate. A circuit can also be provided to condition the indicating appliance circuit to operate despite of an open circuit.

The parallel supervision circuit is similar to the end-of-line device circuit, except that the panel contains additional circuitry in the form of a bridge circuit. (See Figure 6-8.) After all the appliances are connected, the bridge circuit is balanced at the panel. Subsequently, an open circuit in the wiring, or the coil or contacts of a single appliance, unbalances the circuit and causes a trouble signal. In addition to supervising the installation wiring, use of the parallel circuit allows all of the indicating appliance coils and contacts to be supervised.

• *Combination Initiating and Indicating Circuits:* A hybrid circuit that combines both alarm initiating and alarm indicating on a single pair of wires along with supervision of the field wiring is shown in Figure 6-9.

In the normal supervisory condition, the circuit polarity is such that there is no current flowing through the alarm indicating appliances. [Terminal No. 1 on the control panel would be minus (-) and terminal

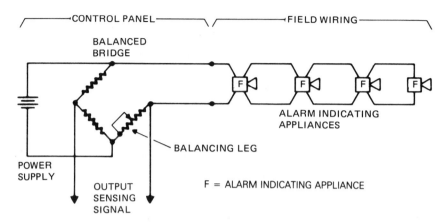

Figure 6-8. Indicating appliance parallel supervision circuit.

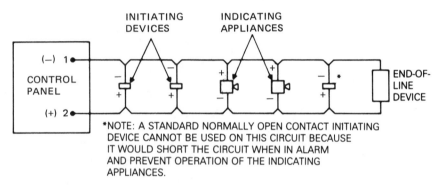

Figure 6-9. Combination alarm initiating and indicating circuit.

2 is plus (+).] A diode in each alarm indicating appliance blocks the flow of current. In the alarm condition, the polarity is reversed as shown in Figure 6-7, and current flows through the alarm indicating appliance.

The alarm initiating device must be polarized or contain a current-limiting means, or both, so the device does not short the circuit when in the alarm condition. Alarm initiating devices must also withstand the full circuit alarm voltage.

• *Signaling Line Circuits:* Signals are transmitted over signaling line circuits from the protected premises to a central supervisory station in all fire alarm signaling systems, except local systems. Although these circuits can take different forms depending upon the type of system, all should be supervised (monitored) in some manner.

The McCulloh circuit, one of the oldest forms of a signaling line circuit, is a coded circuit that is still used in many central station systems. These circuits allow the system to receive an alarm even though the fire alarm circuit may be open because of a fault in the box circuit wiring. An alarm can be received via the remaining good leg of the circuit and ground return. Response to the fault at the control panel can be either automatic or manual. When response is automatic, relays perform the conditioning function; when manual, a switch must be operated. The automatic grounding-type system is free of circuit grounds until a break occurs. A relay is then deenergized and both sides of the line are tied together (via the relay contacts) at the positive terminal of the power supply. Concurrently, the negative side of the power supply is grounded. In the event of an alarm under these conditions, the fire alarm box mechanism will transmit its signal from the positive line via the signaling contacts, to the signal-responsive devices at the control panel, through ground return to the negative power source. Both boxes and the central control panel must be designed for and compatible with this type of operation under an open circuit fault.

In the basic circuit operation, the circuit operates from a grounded positive battery supply. (See Figure 6-10.) The ungrounded negative side feeds through the right-hand circuit relay R_R (one leg of a closed series loop), out through the signal line to the alarm box signaling contacts, back through the left-hand ground relay R_L, and returns to ground. This feed forms the normally closed box circuit, which exists during normal operation.

Under an open fault condition, both the right- and left-hand circuit relays, R_R and R_L, are released, initiating a trouble signal. (See Figure 6-11.) The circuit at this moment is either automatically or manually switched to feed battery out through both relays to the open fault. This forms two normally open circuits that will receive signals from the ground contact of the fire alarm boxes or transmitters.

Under a ground fault condition, the left-hand relay (R_L) releases, initiating a trouble signal. (See Figure 6-12.) The system is then conditioned by switching S1 to feed to minus (-) potential through both relays, out through the circuit to the ground fault, and back to the battery through the grounded plus (+) side. Upon operation of a fire alarm box, the coded pulse is transmitted each time the brush assembly opens the line. Each box or transmitter is equipped with a set of contacts known as a McCulloh brush, which consists of several contact fin-

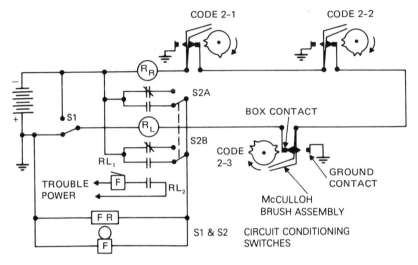

Figure 6-10. McCulloh circuit normal supervisory condition.

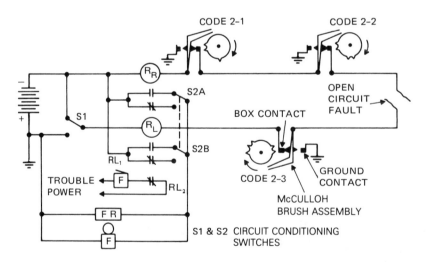

Figure 6-11. McCulloh circuit with open fault condition.

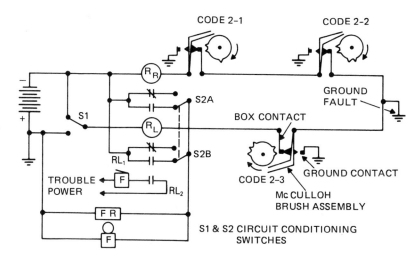

McCULLOH CIRCUIT
GROUND FAULT CONDITION-POWER ON
RELAYS R$_R$ & R$_L$ ENERGIZED
S1 TRANSFERRED

F = ALARM INDICATING APPLIANCE

Figure 6-12. McCulloh circuit with ground fault condition.

gers held together to form a closed and an open circuit. Two of the contact fingers are electrically common but are arranged to flex separately.

For proper performance, the opening and grounding functions of the signaling line circuit must be independent of each other. The open and ground contact blades are staggered so that each code wheel tooth will open, close, and then ground the line. If the line does not close before the ground contact is made, the ground signal will not return both ways, and the signal will be lost under some conditions. Signaling line circuits require the McCulloh brush-type contact action at the fire alarm transmitter to be compatible with its associated control unit.

REMOTE STATION PROTECTIVE SIGNALING SYSTEM

Another method of connecting a fire alarm system directly to the municipal communication center is found in a remote station protective signaling system, which does not use municipal fire alarm circuits. In this system, a separate pair of telephone wires is leased from the telephone company between each property and the municipal communica-

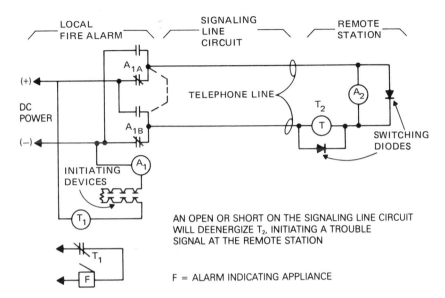

Figure 6-13. Remote station protective signaling system with reverse polarity circuit.

tion center. Generally, these systems are noncoded, and actuation of an individual pilot light identifies the property where the alarm originated. (See Figure 6-13 and Figure 6-14.)

The system in Figure 6-13 is shown with power on, both "A" relays deenergized, "T" relays energized, and in supervisory (monitored) condition. Operation of an initiating device energizes A_1, switching A_{1A} and A_{1B}, which causes a reverse flow of current through the remote station and energizes the system. An open or short on the signaling line circuit will deenergize T_2, initiating a trouble signal at the remote station.

There are two types of noncoded remote station protective signaling systems. The first is the reverse polarity type in which power for the leased circuit originates at the fire alarm control on the protected property. In the event of an alarm, the polarity of the leased line is reversed, e.g., through a relay, so that a relay is energized in a receiver at the municipal communication center. A light goes on, identifying the location from which the alarm originated and actuating an alarm-sounding appliance. Lines are electrically supervised.

The second type of noncoded system utilizes differential current relays or equivalent circuits. A small supervisory current flows through the leased wire and an end-of-line resistor at the protected property. When an alarm is actuated, a relay contact shorts out the end-of-line

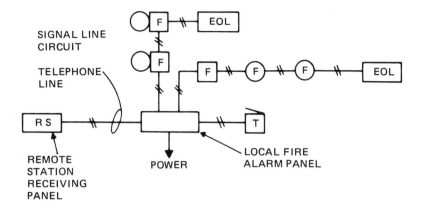

SIGNAL LINE
CIRCUIT

TELEPHONE
LINE

REMOTE
STATION
RECEIVING
PANEL

POWER

LOCAL FIRE
ALARM PANEL

EOL = END-OF-LINE DEVICE
  F = ALARM INDICATING APPLIANCE
  T = TROUBLE SIGNAL

Figure 6-14. Remote station protective signaling system riser diagram.

resistor, causing the current to increase sufficiently to pick up or energize an alarm relay at the municipal communication center. This, in
turn, lights a pilot light and sounds the alarm. This system can be powered from either end of the line. In essence, the remote station is the
control panel and the signaling line circuit is the field wiring between
terminals 1 and 2 and the first initiating device. The initiating devices
are located in the protected building.

ADDRESSABLE ALARM INITIATING DEVICES

Another method of monitoring the interconnection of alarm initiating
devices is with a fire alarm system that is designed to use addressable
detectors. In this system, the central control unit periodically sends an
individually coded signal to each addressable alarm initiating device,
receiving a signal response in return. A trouble signal is initiated if the
control unit either does not receive a response or (in some systems)
receives a return signal indicating that a smoke detector, for example,
exceeded its calibrated limit. Since the devices are addressed individually, there is no need for an end-of-line device or a supervisory current.

When certain addressable systems are used, the alarm initiating circuit can be wired with branch circuits to other addressable devices. The
end-of-line device would be omitted here, and fewer wires are required
in riser Nos. 1 and 2.

ANALOG ALARM INITIATING DEVICES

Analog smoke detectors provide three main features: (1) identification of the fire area, (2) sensitivity setting of the devices, and (3) verification times for each device in the system. The analog value from each smoke detector is the percent of smoke obscuration detected, and is continuously stored and evaluated by the controller. When the value continuously, but slowly increases over an extended period of time, an alert message is generated and displayed. This indicates the need for detector cleaning and maintenance, thereby reducing false alarms due to dirty detectors. When the analog value from the detector increases rapidly, the device, if identified for alarm initiating, will wait until the predetermined verification time has expired. The device will then determine the new analog value from the detector, and place the system into alarm if a sufficient value for alarm condition is present.

NONSUPERVISED PARTS OF FIRE ALARM SYSTEMS

The internal components of alarm initiating devices and alarm indicating appliances are not required to be supervised, although the power supply and the installation wiring that connect the alarm initiating devices and the alarm indicating appliances to the control panel should be supervised.

Testing laboratories review design and performance of the internal components of a fire alarm system under test conditions to determine whether they should be listed for fire alarm service. This ensures some level of reliability. Even with this review, the known failure rate of electronic components necessitates regular testing of a fire alarm system. Testing a fire alarm system is the only way to determine if the system has failed. More information on system testing can be found in Chapter 14 "Testing."

BIBLIOGRAPHY

NFPA Codes, Standards, Recommended Practices, and Manuals. (See the latest *NFPA Catalog* for availability of current editions of the following documents.)

NFPA 72, *National Fire Alarm Code.*

7

Signal Notification

INTRODUCTION

Although it is the last function of the fire alarm signaling process, signal notification, i.e., notifying occupants of a facility that a fire condition exists, is the most important life safety feature of a fire alarm signaling system. The signal notification function also allows certain emergency control measures, e.g., elevator recall, closing of smoke dampers, activation of emergency lighting, etc., to be taken prior to fire department arrival at the scene.

Various types of audible and visible signals, as well as an emergency voice/alarm communication system, can be used as a notification method. The fire alarm signaling system designer must ensure that any audible and/or visible notification signal, and the appliances used to sustain them, are appropriate for the facility and the occupants protected.

This chapter describes notification signals, including types classification; combination systems; use in other signaling systems; and operation, including audible and visible signal appliances. It also addresses the physically challenged occupant and presents appropriate signal design parameters. Signal notification appliance options are detailed. Emergency voice/alarm communication systems, emergency controls, and special considerations for high-rise buildings are also described. For related information see Chapter 4, "Signal Transmission"; Chapter 6, "Electrical Supervision"; and Chapter 13, "Installation."

SIGNAL NOTIFICATION TYPES

Signal notification is accomplished through two methods: (1) audible and (2) visible. Audible signal appliances employed in fire alarm systems for alarm, supervisory, and trouble signals include electro-mechanical appliances, such as bells, horns, buzzers, chimes, and sirens; integral electronic tone-generating appliances, i.e., tone generator, amplifier, and speaker in a common housing; and speakers energized by a remote amplifier source. The suitability of the type of appliance for a particular application and its location depends on the applicable local codes. If the system is a local alarm system, it is recommended that at least one audible fire alarm signal be located outside the building to alert persons in the vicinity who, in turn, could summon fire fighting assistance.

Visible signal appliances rely on two methods: (1) direct viewing of an appliance light source or (2) use of reflected light.

Evacuation Signals

Fire alarm systems provided for evacuation of occupants should have one or more audible signaling appliances approved for this purpose on each floor of the building, and so located that their operation will be heard clearly regardless of the maximum noise level obtained from machinery or other equipment under normal occupancy conditions. (Each section of a floor divided by a fire wall can be considered as a separate floor for the purpose of this protection.) To ensure that audible evacuation signals are clearly heard, it is recommended that they have a sound level of at least 15 dBA above the equivalent sound level or 5 dBA above the maximum sound level having a duration of at least 60 seconds (whichever is greater) measured 5 ft (1.5 m) above the floor in the occupiable area. The equivalent sound level is the mean square, A-weighted sound pressure measured over a 24-hour period. (The "A" is a weighted scale related to the response for normal human hearing.)

Distinctive Signals

Fire alarm signals should be distinctive in sound from other signals and this sound used for no other purpose. Audible signal appliances for a fire alarm system should produce signals that are distinctive from other similar appliances used for other purposes in the same area.

The recommended fire alarm evacuation signal is a uniform Code 3 temporal pattern, using any appropriate sound, keyed $\frac{1}{2}$ to 1 second "ON," $\frac{1}{2}$ second "OFF," $\frac{1}{2}$ to 1 second "ON," $\frac{1}{2}$ second "OFF," $\frac{1}{2}$ to 1 second "ON," and $2\frac{1}{2}$ seconds "OFF," with timing tolerances of $\pm$ 25 percent, repeated for not less than 3 minutes. Manual interruption of the repetition time is permitted by some local codes.

The recommended standard fire alarm evacuation signal is intended for use only as an evacuation signal. A different sounding, separate signal should be used to relocate the occupants from the affected area to a safe area within the building, or for their protection in place (e.g., in high-rise buildings, health-care facilities, penal institutions, etc.).

Supervisory signals should also be distinctive in sound from other signals and used only to indicate a trouble condition. Fire alarm signals should take precedence over all other signals.

Trouble Signals

Trouble signals should be distinctive from alarm signals and indicated by the operation of a sounding appliance. If an intermittent signal is used, it should sound at least once every 10 seconds with a minimum time duration of 0.5 second. An audible trouble signal can be common to several supervised circuits. The trouble signal(s) should be located in an area where it is likely to be heard, as designated by the authority having jurisdiction.

Visual Zone Alarm Indication

When required by local code, location of an initiating device should be visually indicated by building, floor, fire zone, or other approved subdivision by annunciation, printout, or other approved means. The visual indication should not be cancelled when an audible alarm silencing switch is operated.

Notification of occupants is of prime importance during a fire emergency, but the need to notify the fire department is equally important. If NFPA *101*, *Life Safety Code*, is used by the local authority and the local code requires fire department notification, this signal should automatically be transmitted by an auxiliary signaling system, a central station, remote station signaling system, or a proprietary system operator. If a fire alarm signaling system is not used, some type of plan should be provided to notify the fire department of a fire emergency.

CLASSIFICATION OF NOTIFICATION SIGNALS

For the purpose of this text, notification signals for protective signaling systems are classified as noncoded, coded or textual, and with public or private operating modes.

Noncoded, Coded, or Textual Signals

Noncoded signals are audible or visible signals and convey one discrete bit or unit of information; a coded signal conveys several bits or units of information. A textual signal conveys a stream of information. For example, one stroke of an impact-type appliance is a noncoded signal, as is the continuous operation of an appliance that is continuously energized (or interrupted at a continuous uniform rate). Examples of coded signals are numbered strokes of an impact-type appliance or numbered flashes of a visible appliance. A voice message is one example of an audible textual signal.

Public and Private Operating Modes

Notification signals can have two operating modes: (1) public or (2) private. With a public operating mode, the inhabitants of the protected area receive the audible or visible signal from the protective signaling system. A private operating mode restricts the signal to the persons concerned with initiating and carrying out emergency actions, e.g., a private fire brigade, in the protected area.

Audible and Visible Characteristics

Audible signals intended for operation in the public mode should have a sound level of not less than 75 dBA at 10 ft (3 m) or more than 130 dBA at the minimum hearing distance from the audible appliance. The sound level range for private mode audible signals should not be less than 45 dBA at 10 ft (3 m) or more than 130 dBA at the minimum hearing distance from the audible appliance.

The sound level of an installed signal should be adequate for its intended function and should meet the requirements of either the local authority or applicable standards. The average ambient sound level for the various occupancies as listed in Table 7-1 can be used for design guidance.

There are two primary visible signaling methods: (1) direct and (2) indirect. For a direct visible signaling method, the message of notification of an emergency condition is conveyed by direct viewing of the illuminating element within or attached to the signaling appliance. The indirect method illuminates the area surrounding the visible signaling appliance.

SIGNAL NOTIFICATION IN COMBINATION SYSTEMS

Combination systems consist of equipment used for fire alarm, sprinkler supervisory or watch service, and for other signaling systems such as

TABLE 7-1. Average Ambient Sound Levels

Locations	dBA
Business Occupancies	45
Educational Occupancies	45
Industrial Occupancies	80
Institutional Occupancies	50
Mercantile Occupancies	40
Piers and Water-Surrounded Structures	40
Places of Assembly	40
Residential Occupancies	35
Storage Occupancies	30
High-Density Urban Thoroughfares	70
Medium-Density Urban Thoroughfares	55
Rural and Suburban Thoroughfares	40
Tower Occupancies	35
Underground Structures and Windowless Buildings	40
Vehicles and Vessels	50

burglar alarm, voice paging or music program systems, or a coded paging system. "Combination" also applies to methods of using circuit wiring common to both types of system.

Common Wiring

Equipment not used in fire alarm signaling systems should be connected to common wiring in a combination system so that short or open circuits, or grounds in this equipment or between this equipment and the signaling system wiring, will not interfere with either the supervision of the fire system or prevent alarm or supervisory signal transmission.

Use of Speakers

Speakers can provide the audible alarm signal for local fire alarm or sprinkler alarm systems. They should be used for emergency voice/alarm communication systems and other emergency purposes only, unless other uses are approved by the local authority. Selective paging is usually allowed where the fire command station is constantly attended by a trained operator.

In general, speakers used in fire alarm signaling systems for tone signals and/or emergency voice communications should have a frequency response and power rating suitable for the application. Speaker materials that are subject to moisture absorption should be suitably protected.

In a combination system, the fire alarm signal should be clearly recognizable and take precedence over any other signal, even though a nonfire alarm signal is initiated first.

SIGNAL NOTIFICATION IN OTHER SIGNALING SYSTEMS

Notification specifications differ where a signaling system is not a local protective signaling system. Signal notification should be located at the central supervising station to provide signal notification for evacuation of occupants or signals directing aid to the location of an emergency. Suitability and location of indicating appliances for alarm, supervisory, and trouble signals should be determined by the authority having jurisdiction.

Signal notification of the section of the building from where the signal originated should be designated at either the central supervising station or at the building protected. Indicating appliances to pinpoint the location of a fire should be acceptable to the local authority.

Although the use of audible alarm signals (other than one at the central supervising station) is not required in other than local systems, any desired evacuation signal should be compatibly designed into the system. As for any system, trouble signals should be distinctive from other audible signals.

SIGNAL NOTIFICATION OPERATION

Closing of a smoke damper in an air-handling duct by the signal initiation from a smoke detector within that duct is an automatic action; it results from an electrical or mechanical extension of the fire alarm system.

The manual action, however, i.e., the evacuation of a building occupant by the signal initiation from a manual fire alarm box, is physiological. This action depends upon human response as a result of the sense of the signal and the interpretation of its meaning. Limitations of human response must be addressed for practical implementation of manual actions. Current notification appliances consider audibility and visibility only out of the five human senses.

Because a fire alarm system must operate before and during a fire emergency, signal notification appliances must operate at elevated tem peratures. Fire resistance is usually a concern only with public mode appliances, since private mode operation is usually protected by fire barriers.

Audible Signal Characteristics

Audibility of a signal notification appliance must be considered for the particular place where the appliance is installed. A gong in a recessed enclosure with a grille is less audible than a surface-mounted gong. Heavy-carpeted floors, fabric walls, and acoustic ceilings are sound deadeners that must also be considered.

The audible signal notification should be approximately thirty times greater than "ambient" or background noise volume. Expressed in decibels (dB), a logarithmic function, this amounts to about 15 dBA above ambient.

The signal for the public mode of operation should have a sound output level of no less than 75 dBA where contributions from signal reflection are eliminated. A minimum ambient level of 85 dBA for mechanical equipment rooms is recommended. Since private mode operation is in a controlled environment, an ambient level would probably be no greater than 35 dBA.

Design of Audible Appliances: Basic criteria for the design of audible fire alarm devices can be found in NFPA 72, *National Fire Alarm Code*. However, design methodology is not addressed in NFPA 72. One method that provides a good estimate for the effective positioning of devices for various fire alerting systems can be found in *The SFPE Handbook of Fire Protection Engineering*. This method is based upon a 1981 British study entitled *Locating Fire Alarm Sounders for Audibility*, written by H. Butler, A. Bowyer, and J. Kew.

The four main parameters required for the location of audible fire alarm devices are the following:

1. Sound characteristics of the floor, walls, and ceiling;
2. Sound pressure level of the device (based upon manufacturer's data);
3. Number of directions that the sound can propagate; and
4. Distance from the source device.

The devices must be designed and located to meet the worst-case condition, i.e., achieving the minimum required sound pressure level in the farthest remote room.

Audible appliances for protective signaling systems should be listed, and protected against effects of corrosion, temperature, humidity, and physical damage.

Audible Signal Location: Signal notification appliances operated in the public mode should be located so the sound pressure level is not less than 15 dBA above ambient. The sound pressure level at the minimum hearing distance from the appliance is determined by the threshold of pain (130 dBA), i.e., the level where sound may cause physical damage to human hearing. For private mode operation, the appliance should produce no less than 10 dBA above ambient over the private mode area, which is usually a building control room.

Where ceiling height permits, wall-mounted appliances should have their tops at heights above the finished floors of not less than 90 in. (2.30 m) and below the finished ceilings of not less than 6 in. (0.15 m). This does not preclude ceiling-mounted or recessed appliances.

Visible Signal Characteristics

There are two types of visible signal notification: (1) when the sole means of notification is by direct viewing of its appliance light source or (2) by reflected light within the room. Since direct appliances can be used under a variety of conditions, the boundary limitations of the signal should be determined to permit maximum versatility in equipment design and application. The signal can be defined in terms of its effective intensity (I_{eff}) by using the following equation:

$$I_{eff} = \frac{\int_{t_1}^{t_2} I(t)dt}{t_2 - t_1 + 0.2}$$

where
I(t) is the instantaneous intensity expressed in candela,
t is in seconds, and
t_1 and t_2 are chosen to maximize the effective intensity.

The boundary conditions of the signal are $A \leq I_{eff} \leq B$. A and B can be defined as follows:

1. Where the average illuminance is less than 5 lumens per sq ft (0.093 m²), the effective intensity is between 1.5 and 15 candela;

2. Where the average illuminance (without motion present) is greater than 5 lumens per sq ft (0.093 m²), the effective intensity is between 15 and 150 candela; and
3. Where the average illuminance (with motion present) is greater than 20 lumens per sq ft (0.093 m²), the effective intensity is between 100 and 1,000 candela.

The direct signal appliance is dependent: (1) on the viewer's line of sight and (2) on the environment. The signal within the defined visible range should have the following directional characteristics:

1. Through ± 85 degrees from the nominal signal direction and not less than 10 percent from 85 to 90 degrees,
2. For viewing only below the horizontal, through +10 degrees to -70 degrees and not less than 10 percent from -70 degrees to -85 degrees, and
3. For viewing above and below the horizontal, through +45 degrees to -70 degrees and not less than 10 percent from +45 degrees to -85 degrees.

Once the ambient illuminance has been determined, the average effective illuminance (E_{eff}) can be determined by using the physical parameters of the environment. For the standard nominal work plane 30 in. (0.76 m) above the floor:

$$E_{eff} = \frac{\int_{t_1}^{t_2} E(t)dt}{t_2 - t_1 + 0.2}$$

where
E(t) is the instantaneous illuminance expressed in lumens per sq ft,
t is in seconds, and
t_1 and t_2 are chosen to maximize the effective illuminance.

When E_a is the steady-state spatially averaged work plane illuminance in the room, then

$$0.02\, E_a \leq E_{eff} \leq 0.1\, E_a$$

Indirect signal appliances should not be used for environments with an average illuminance of less than 1.0 lumens per sq ft (0.093 m²). The direct signal appliance should be used in dark environments, such as those found in theatres and cocktail lounges.

Pulse Requirements: The repetition and duration of light pulses are an important design feature in visible signal notification appliances.

Pulse repetition needs to be limited to a small range, and the upper limit of the range should not appear to be a flickering light. In the lower limit of the range, the visible signal should be recognized as a warning signal. Since the indirect visible signal tends to modulate the environmental light, a higher upper limit can be tolerated for the direct visible signal. Consistent pulse repetition is desirable, since this feature principally identifies the warning. The range for pulse repetition should be within a flash every 3 seconds (for indirect signal) to a maximum of 3 flashes per second (for direct signal).

Pulse duration is the time interval between the initial and final points of 10 percent of the maximum signal time. The duty cycle is the ratio of the pulse duration time to the time between the initial points of two consecutive pulses. Optimally, a principal pulse can be made from a string of pulses, provided the repetition of this string exceeds 30 pulses per second. This allows a system to use a string of pulses for practical control reasons. Maximum pulse duration and duty cycle should be 0.2 of 1 second and 40 percent, respectively.

Color Requirements: The visible signal source color should be white. The white spectral limit is defined on a chromaticity diagram. (See Figure 7-1.) Available light sources best suited for use as visible signals are wide-band radiators, such as xenon discharge lamps. Although other colors can be formed by using filters, this adds unnecessarily to power needs. The pulsing nature of the visible signal is the primary form of identification, not its color. Local jurisdictions may identify yellow (amber) or red as colors to be used in signaling a fire emergency. If this is the case, these colors must be identified in protected areas as only serving that purpose.

Visible Signal Location: The location of a direct signal appliance should be determined by the needs of large enclosures that are not rectangular, multistory (such as atria), or contain spatial obstructions such as manufacturing areas and libraries. Direct signal appliances should be located in these large enclosures as described herein.

The possibility of either temporary or permanent blockage of the appliance by portable objects should be minimized. This can be accomplished with a light source opening that is no less than 6.5 ft (2 m) above the highest floor level in the room.

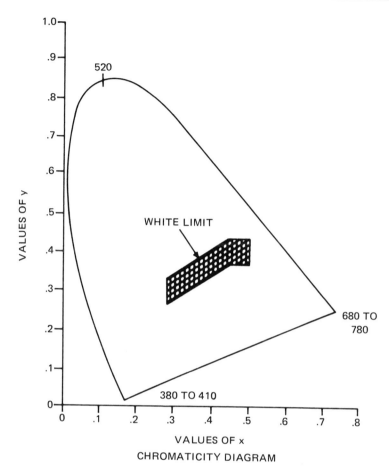

Figure 7-1. Chromaticity diagram for definition of the white spectral limit.

The power level of the appliance should be specified for the particular area covered. An appliance designed to provide notification for a large area could be objectionable to occupants of a smaller enclosed space. There should be a minimum of one additional appliance for each 7,500 sq ft (697 m²) of floor area in excess of 10,000 sq ft (929 m²). At least one appliance should be directly visible for 95 percent of the occupied floor area as measured in the protection plan. It is not practical to provide a signaling appliance that is within lines of sight of all building occupants. Multiple appliances can be located so that at least two appliances are separated by a minimum of 135 degrees in plan view for 75 percent of the occupied floor area. An appliance should be within +60 degrees and -30 degrees (vertical angles) over 75 percent of the floor plan area that the appliance covers. Considering the mobility of the lines of sight, reflections from objects in the room, and interaction between occupants, it is not necessary for a direct signal appliance to be within the fields of view at all times.

The location of the indirect signal appliance should be determined from the average effective illuminance using the space and the visible characteristics of the appliance. The location parameters can be calculated using space shape and dimensions, surface reflectances, and appliance height and output. (See Figure 7-2.) The exact mathematical equation to locate an indirect appliance solution can be approximated as follows:

1. Divide the space into a room cavity (r) between the working plane [30 in. (0.76 m) above the floor] and the height above the floor for the appliances, and a ceiling cavity (c) between the height above the floor for the appliances and the ceiling. The average ceiling height should be used when the ceiling is not horizontal.

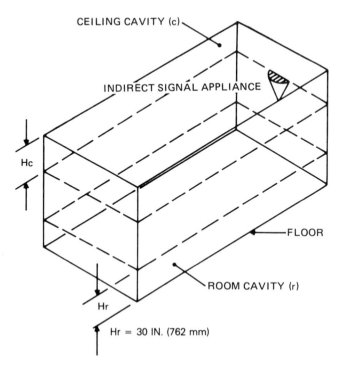

Figure 7-2. Indirect signal appliance space.

2. Determine the shape factors (M_r and M_c) which are equal to the height of the cavity multiplied by the perimeter of the cavity divided by the area of the cavity. This is for any space without a reentrant perimeter. (Small alcoves, etc., should be disregarded.) "L"-shaped spaces may be divided into two rectangular spaces. The shape factors simplify, for a rectangular room, to twice the height of the cavity multiplied by the sum of the width and length, divided by the product of the width and length.

3. The effective reflectance of the ceiling cavity at its lower boundary should be determined by calculating

$$\frac{\text{ceiling reflectance}}{1 + 0.6\, M_c}.$$

When the ceiling does not have uniform reflectance, the ceiling reflectance should be its spatially weighted average.

4. The flux transfer factor from the appliance to the work plane should be determined by calculating

$$\frac{2 \times \text{ceiling cavity reflectance}}{1 + 0.6\, M_r}.$$

5. The average effective illuminance is the effective luminous flux emitted by the appliances, multiplied by the flux transfer factor divided by the floor area. The effective luminous flux emitted by the appliances (ϕ_{eff}) is a parameter determined by their manufacturer in accordance with the equation

$$\phi_{\text{eff}} = \frac{\int_{t_1}^{t_2} \phi(t)\,dt}{t_2 - t_1 + 0.2}.$$

where
$\phi(t)$ is the instantaneous spatially total luminous flux in lumens emitted by the appliances,
t is in seconds, and
t_1 and t_2 are chosen to maximize the luminous flux.

6. These relations can determine the number and ratings of the appliances needed or evaluate present appliances. They should not be used for very dark space surfaces such as ceilings with less than 25 percent reflectance or walls with less than 20 percent reflectance. Spaces with very dark surfaces should be provided with direct signaling appliances.

The means used for locating the indirect signal appliance indicates that ceiling reflectance has a much greater effect on the flux transfer versus wall reflectance. Generally, ceiling reflectance range exceeds that for wall reflectance. The factor of wall reflectance may be eliminated from the calculations for locating the appliances. The error range introduced, as compared to the exact mathematical solution, is shown in Table 7-2.

TABLE 7-2. Error Ranges

| | *Extreme Error Limits* | | |
Space Factors	*Effective Ceiling Reflectance*	*Flux Transfer Factor*	*Total*
Common Range:			
$M_r \leq 2$	+25% to -40%	±20%	±50%
$M_c \leq 0.8$			
Extended Range:			+150%
$M \leq 4$	+70% to -60%	+50% to -20%	to
$M_c \leq 2$			-70%
Range:			
Ceiling Reflectance	90% to 25%	90% to 20%	
Wall Reflectance	80% to 20%	60% to 20%	

Extreme errors occur for the least likely combinations of ceiling reflectance, wall reflectance, and room dimensions. These errors will be generally much less than the extremes. For the common range of space factors, the maximum error will not exceed a 2:1 ratio; and for the extended range of space factors, maximum error will not exceed a 3:1 ratio.

The following additional points should be considered in locating the indirect signal appliance:

1. No direct viewing of the lamp should be permitted from below the horizontal plane. No optical element (either lens or reflector)

should be optically active externally at angles below the horizontal plane. This permits part of an optical component within the field of view, provided it is not directing light to that region.

2. The light from the appliance should be reasonably well distributed over the ceiling and the upper walls. There should also not be an extremely bright area adjacent to the appliance. Since the appliance is likely to be ineffective in the more distant parts of the protected space, explicit design limitations on the three-dimensional intensity distribution should ensure good light distribution.

3. For wall-mounted appliances, no more than 15 percent of the output should be directly incident on the plane of the wall. No more than 25 percent of appliance output should be incident on the ceiling within 45 degrees of the vertical through the appliance. A minimum of 40 percent of the output should be between the horizontal appliance plane and 30 degrees above. (See Figure 7-3.)

4. The vertical angle below which 10 percent of the appliance output occurs should be established. The maximum horizontal distance of coverage should not exceed four times the distance at which a line

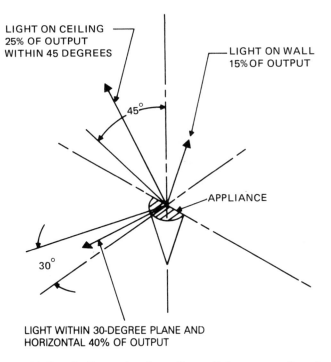

LIGHT ON CEILING
25% OF OUTPUT
WITHIN 45 DEGREES

LIGHT ON WALL
15% OF OUTPUT

45°

APPLIANCE

30°

LIGHT WITHIN 30-DEGREE PLANE AND
HORIZONTAL 40% OF OUTPUT

Figure 7-3. Indirect signal appliance light output directions.

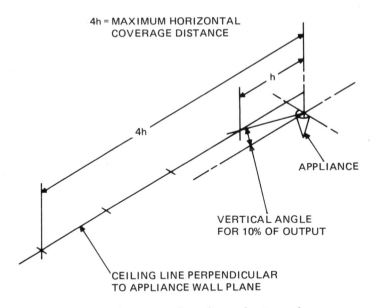

Figure 7-4. Indirect signal appliance horizontal coverage.

through the appliance at this angle intersects the ceiling. For example, a manufacturer might express an appliance rate for maximum coverage distance as a multiple of the appliance-to-ceiling distance. (See Figure 7-4.)

Visible Signaling Appliances: As an alternative to audible appliances, NFPA 72, *National Fire Alarm Code*, permits visible signaling appliances. These appliances can provide hearing-impaired persons with equivalent fire emergency notification that hearing persons can receive through audible signaling appliances.

The minimum criteria to ensure that visible signaling devices can be seen by building occupants should satisfy the following for direct primary visualization.

Average Illuminance in Protected Area	Effective Intensity Rating Pulsing Light
< 5 lumens/ft^2	0.15 to 15 candela
> 5 lumens/ft^2	15 to 150 candela
> 20 lumens/ft^2 + motion	100 to 1000 candela

The minimum criteria to ensure that the visible signaling appliance can be seen by the building occupants without causing eye damage should satisfy the following for indirect primary visualization.

1. The average effective illuminance of the pulsing signal at a standard nominal work plane 30 in. (0.76 m) above the floor should be a minimum of one-tenth the steady-state spatially averaged work plane illuminance at the standard nominal plane in the room.

2. For an appliance mounted on or adjacent to a wall, it is recommended that not more than 15 percent of its output be directly incident on the plane of the wall and not more than 25 percent be incident on the ceiling within 45 degrees of the vertical through the appliance. A minimum of 40 percent of the output should be between the horizontal and +30 degrees from the horizontal.

3. Direct viewing of the lamp must not be permitted from below the horizontal plane. This will prevent eye damage. No optical element, either lens or reflector, should be optically active externally at angles below the horizontal. Part of the optical component within the field of view should be permitted, provided it is not directing light to that region.

A NFPA 72, *National Fire Alarm Code*, exception to the above is that combination indirect- and direct-viewing appliances are permitted if the direct viewing component does not exceed an effective intensity of 1000 candela.

NFPA 72, *National Fire Alarm Code*, recognizes two methods of primary visible signaling: (1) direct detection and (2) indirect detection. Direct detection methods include a visible signaling method in which the message of notification of an emergency condition is conveyed by direct viewing of the illuminating element within or attached to the signaling appliance. The indirect detection method is one in which the message of notification of an emergency condition is conveyed by means of illumination of the area surrounding the visible signaling appliance.

It is important to remember that, when locating visible alarm indicating appliances, the main objective is to notify hearing-impaired occupants of a fire emergency in a way equivalent to that provided hearing persons by audible signaling appliances.

As a standard, NFPA 72, *National Fire Alarm Code*, estimates the average illuminance for various activities, as listed in the following chart:

Type of Activity	Average Illuminance (Lumens/ft^2)
Public spaces with dark surroundings	2.5
Simple orientation for short temporary visits	5-10
Working spaces where visual tasks are only performed occasionally	10-20
Performance of visual tasks of high contrast or large size	20-50
Performance of visual tasks of medium contrast or small size	50-100
Performance of visual tasks of low contrast or very small size	100-200

Since visible alarm indicating appliances must be seen to be effective, it is generally necessary to use more appliances than audible indicating appliances. Reflective techniques can be used to reduce the number of devices.

Supplementary Visible Signals: NFPA 72, *National Fire Alarm Code*, recognizes a supplementary visible signal that is utilized to augment an audible or primary visible signal. The supplementary visible signal need not meet the light intensity recommendations of primary visible signals. One of the key design parameters is that the combined intensity of both the primary and supplementary appliances should not exceed the upper bounds for the primary visible signal.

THE PHYSICALLY OR MENTALLY CHALLENGED

The physically or mentally challenged population includes the hearing- and sight-impaired, some elderly persons, people with mental deficiencies or incapacities, etc. This population must be considered when an audible or visible signal and appliances are specified for a building.

The use of an encompassing pulsing audible or visible signal must be examined to determine if it will cause distress or confusion to some occupants. These consequences are important, since the pulsing audible or visible signal is used to alert potentially hazardous conditions. Further, pulsing audible and visible signal rates can have hurtful effects for persons with epilepsy.

Any negative effect caused to the physically or mentally challenged occupant by signal notification for fire alarm systems should be examined for appropriate modifications of the signals.

Americans with Disabilities Act of 1990

Federal Register rules and regulations for the 1990 Americans with Disabilities Act (ADA) delineate specific requirements for fire alarm systems pertinent to physically or mentally challenged building occupants.

The ADA requires both visual and audible alarm appliances for fire alarm systems. All occupancies are included in the law except medical care facilities. Fire alarm systems are not required by the ADA, but once a fire alarm system is planned for a facility, ADA requirements must be incorporated into the new system.

ADA requires that audible alarms have sounding levels that are 15 dBA above ambient noise levels, or levels that are 5 dBA above the maximum sound level for a 60-second duration. The regulations establish maximum alarm levels at 120 dBA for audible signals. Audible emergency signals must have an intensity and frequency that can attract the attention of individuals who have partial hearing loss. The Act indicates that people over 60 years of age have difficulty perceiving frequencies higher than 10,000 Hz. The best appliances for the partially hearing impaired are single-stroke bells, high-low alarm signals, and the fast whoop signals. The regulations suggest signals with a sound characterized by three or four clear tones without a compromising "noise" in between.

The ADA parameters differ from previous criteria established in ANSI A17.1, *Safety Code for Elevators and Escalators*, which is an accessibility standard used currently in the public sector. The ADA requirements are significantly different from the recent edition of ANSI A17.1, both in terms of light intensity and lamp distribution. Other differences with ANSI A17.1 include minimum lamp intensities of 185 candela, while ADA has a minimum lamp intensity of 75 candela.

Besides differences with ANSI A17.1, the ADA requirements also differ from those in NFPA 72, *National Fire Alarm Code*. These three documents have conflicts with light levels for visual alarm appliances. These conflicts concern light output levels based upon room size. Besides light intensities, the three documents differ in appliance maximum mounting heights, and flashing rates for strobe lights, which has the potential for causing epileptic seizures. Some of the differences are illustrated in Table 7-3.

TABLE 7-3. Minimum Required Light Output in Candela (cd)*

	NFPA 72			ANSI A117.1 1992	ADA
Maximum Room Size	10 ft Max. Ceiling Height	20 ft Max. Ceiling Height	30 ft Max. Ceiling Height	80 in. to 96 in. Mounting Height	80 in. Max. Mounting Height
20 × 20 ft	15 cd	30 cd	55 cd	15 cd	75 cd
30 × 30 ft	30 cd	45 cd	75 cd	30 cd	75 cd
40 × 40 ft	60 cd	80 cd	115 cd	60 cd	75 cd
50 × 50 ft	95 cd	115 cd	150 cd	95 cd	75 cd

* This table is a partial extraction of the criteria for light levels from NFPA 72, *National Fire Alarm Code*; ANSI A117.1, *Accessable and Usable Buildings and Facilities*; and the *Americans with Disabilities Act.*

For SI Units: 1 in. = 25.4 mm; 1 ft = 0.30 m.

The Act also requires visual alarm signal appliances to be integrated into building or facility fire alarm systems. In essense, the features and positions of visual alarm devices are described in the Act as follows:

1. The lamp must be a xenon strobe type or equivalent.
2. The color must be clear or "normal" white.
3. The maximum pulse duration must be two tenths of one second (0.2 sec), with a maximum duty cycle of 40 percent. The pulse duration is defined as the time interval between initial and final points of 10 percent of maximum signal.
4. The intensity must be a minimum of 75 candela.
5. The flash rate must be a minimum of 1 Hz and maximum of 3 Hz.
6. The appliance must be placed 80 in. (2 m) above the highest floor level within the space or 6 in. (150 mm) below the ceiling, whichever is lower.
7. In general, any room or space required to have a visual signal appliance must have no spot more than 50 ft (15 m) from the signal (in the horizontal plane). In large rooms and spaces exceeding 100 ft (30 m) across, without obstruction 6 ft (1.8 m) above the finish floor, such as auditoriums, devices can be placed around the perimeter, spaced a maximum 100 ft (30 m) apart, in lieu of suspending appliances from the ceiling.
8. No place in common corridors or hallways in which visual alarm signaling appliances are required is allowed to be more than 50 ft (15 m) from the signal.

Residential Alarm Appliances for the Hearing Impaired

In sleeping areas, where visual alarm devices are placed, the signal must be visible in all areas of the unit. In addition, the visual alarm devices must be connected in some way to the building emergency alarm system. For visual alarms to be effective for the hearing impaired, the devices must be located and oriented so that they will spread the signals and reflect the light throughout the space or raise the light level sharply in the space.

In residential-type occupancies for the hearing impaired, visual alarms are not the best method for signaling. Underwriters Laboratories Inc. has performed tests and conducted studies that have determined that to awaken sleepers flashing lights with the intensity of 110 candela were needed. This is over seven times more intense than the normal 15 candela. A more effective way to awaken sleepers is the installation of vibrators in the bedding area to alert sleepers of fire emergencies.

SIGNAL NOTIFICATION APPLIANCE OPTIONS

Audible Alarm Signal Appliances

Bells: Bells can be used for fire alarm signals if they have a distinctive sound and will not be confused with similar audible signals used for other purposes, such as school classroom bells. Bells can be of the single-stroke or vibrating type. Single-stroke bells are used to provide audible coded signals; continuous vibrating types are used primarily for noncoded, continuous-ringing applications. Continuous vibrating bells also can be used to provide coded audible signals.

Bells can be provided with 4-in. through 12-in. (100-mm through 300-mm) gongs in 2-in. (50-mm) increments, although the 6-in. and 10-in. (150-mm and 250-mm) sizes are the most commonly used. Bells with 4-in. (100-mm) gongs are usually reserved for use as trouble signals. Generally, the larger the diameter of the gong, the lower the frequency and the louder the audible signal as expressed in decibels.

Bells are usually of the underdome type and can be mounted on standard conduit boxes. Special boxes and grilles are necessary where bells must be concealed, recessed, and/or mounted flush with the wall.

Horns: Horns can be used where louder and/or more distinctive signals are needed. Horns may require more operating power than bells, and circuits should be electrically compatible where the power source supplies both types of signals.

Horns are usually of the continuous vibrating type and can provide either coded or noncoded audible alarm signals. They may be the surface (grille), flush, semiflush, single projector, double projector, or trumpet type. In very noisy areas, resonating, air-powered, or motor-driven horns are sometimes used because of their inherently high-decibel output. Air-powered horns using valves are controlled by ac- or dc-powered electrical solenoids. Either coded or noncoded operation may be provided.

Motor-driven horns are not practical for providing coded output signals and are more widely used for continuous signals. They are particularly effective under "rolling" conditions where power is periodically applied and removed to vary the motor speed and the sound pitch.

Chimes, Buzzers, and Sirens: Chimes are soft-toned appliances that are used where panic or other undesirable actions might result from the use of loud audible alarm signals. Their use is especially compatible with such areas as nurses' stations in hospitals, whereby only authorized personnel is alerted to a fire emergency. Buzzers are generally used for trouble signals rather than alarm signals and are intended to provide a continuous sound. They are seldom used for coded signals. Sirens are usually motor-driven or electronic appliances and limited to outdoor applications, but are sometimes used in extremely noisy indoor areas.

Speakers: Speakers reproduce electronic signals and can be made to sound like any mechanical signaling device. They are capable of reproducing unique sounds that are not practical on mechanical appliances. They can also be used to provide live or recorded voice instructions. Speakers are either the direct-radiating-cone type or the compression-driver-and-horn type, operated from audio amplifiers delivering standard output line levels of 70.7- or 25-volt rms. The speakers are driven by an electronic tone generator, microphone, tape player, or voice synthesizer and electronic amplifier.

The two most widely used speakers are: (1) the integral type, where the tone generator, amplifier, and speaker are enclosed in common housing; and (2) the remote type, where the speaker is energized from a remotely located generator, amplifier, etc.

Visible Alarm Signal Appliances

Visible alarm signal appliances of the incandescent, strobe, or quartz halogen variety may be used alone or in conjunction with audible alarm appliances.

Visible alarm indicators are effective in high-noise areas, to alert the hearing-impaired occupant, or where attention directed to the light unit can provide additional information. For most applications, visible alarm appliances are most effective when used in a flashing mode. Flashing can be caused by periodically interrupting current to the appliance, rotating a beam of light (as is done on many emergency vehicles), using a strobe technique, or any combination of these methods.

Examples of visible alarm signal appliances are lamp annunciators; strobe lights; and incandescent, solid-state, quartz halogen, and fluorescent lamps. Annunciators can be used with either coded or zoned non-coded systems to indicate the initiating device or zone from which the alarm orginated.

Lamp Annunciators: In lamp annunciators, the zones or areas from which alarms originate are indicated by the lighting of lamps or light-emitting diodes (LEDs). Lamp annunciators can be provided with different colored lenses (usually red for alarm, amber for trouble, and green or white for power-on indications).

Back-lighted annunciators are used to light up information on a window or to illuminate a graphic annunciator. Graphic annunciators have lamps specified in a map or floor plan to provide an easier means of identifying the particular area or zone affected. The area from which the alarm was initiated can be indicated by a "bull's-eye"-type lamp located within the area or by backlighting the entire affected area or zone.

Strobe Lights: Strobe lights operate on the energy discharge principle to produce a high-intensity, short-duration flash. These lights are very efficient, producing, for example, 8,000 candlepower for 1 watt input power. The short bright flash is not only attention getting but effective with low general visibility. Strobe appliances come in a wide range of light intensities. Repetition rates are usually between one and five flashes per second.

Lamps: Incandescent lamp alarm appliances can be used to provide coded or noncoded visible indication and can be used in the continuous or repetitive flashing modes.

Solid-state luminous devices, e.g., a LED lamp, are becoming more prevalent because of their high reliability, relatively low current rating, and availability in various colors. These lamps are inherently low-current devices adaptable to solid-state energizing circuits.

Quartz halogen lamps draw relatively high current but are an efficient source of high-intensity light. They can be used in repetitive or continuous high-intensity operation. Quartz halogen devices are particularly effective as the light source for rotating reflector flashing lamps.

Fluorescent lamps are used primarily as sources of distributed or diffused backlighting in continuous warning signals. When activated, the signals convey a printed warning or direction message. Because of their complex starting circuits and relatively low brilliance, fluorescent signal appliances are usually not used for coded or flashing warning applications.

Combination Audible and Visible Appliances

Audible and visible functions can be combined to produce both sound and light from a single appliance. For example, the sounder can be a horn, bell, or speaker. The light can be incandescent, strobe, or other type, depending on the application.

The visible signal identifies the particular audible alarm appliance that is operating and produces a recognizable alarm when the audible signal may be obscured by an ambient noise level. Also, persons having impaired hearing can see the visible portion of the alarm signals.

Permanent and Print Recorders

Permanent recorders are used on fire alarm systems to indicate the location of an alarm and to provide a permanent record of the time and date an alarm occurred. A recorder that creates a hard-copy record of the alarm location performs the same function as a lamp annunciator.

In many systems, the recorder can be used to record a trouble condition plus other events that are related to the receipt of an alarm. Examples of other events are alarm indicating circuit energization, fan shutdown, door closing, or other life safety functions that may have been initiated either automatically or manually.

Electronic print recorders have replaced old punch-type recorders almost entirely in modern fire alarm systems. Print recorders range from small hand-size units (similar to those found on small printing calculators) to large, sophisticated, high-speed printers found on major computer systems. NFPA 72, *National Fire Alarm Code*, requires two printers, one of which is used for all incoming signals, and the other reserved for fire alarm, supervisory, and trouble signals only when no other visual means of indicating an alarm is provided.

Textual Appliances

Textual audible appliances are similar for the public and private modes and, at present, include the loudspeaker, the telephone, and the radio frequency appliances (radio telephone). Loudspeaker appliance output should be in accordance with the sound output levels previously defined for audible appliances. Since telephone appliances have extensive nonfire alarm system usage, existing definitions for these appliances, issued by the Federal Communications Commission and by the Electronic Industries Association, are usually appropriate. NFPA 72, *National Fire Alarm Code*, specifies requirements for both the loudspeaker and telephone appliances.

The textual visible appliance operates mostly in the private mode and consists of pictorial and alphanumeric displays that can be permanent, with a minimum retrieval time of one year, or may be temporary. Where the textual visible appliance operates in the public mode, e.g., in a building lobby for the fire service, additional requirements are needed to provide operation during fire conditions.

EMERGENCY VOICE/ALARM COMMUNICATION SYSTEMS

Signal notification is increasingly found in the form of an emergency voice/alarm communication system that is added to the building fire alarm system. Emergency voice/alarm communication systems provide dedicated manual and automatic facilities for fire location and control, and transmission of information and instructions concerning a fire alarm emergency to the building occupants, as well as fire department personnel.

Voice/Alarm Signaling Service

Voice/alarm signaling service provides an automatic response upon receipt of a signal that indicates a fire emergency. Subsequent manual control capability of the transmission and audible reproduction of evacuation tone, signals, preannounce alert tone signals and voice directions on a selective and all-call basis (as determined by the authority having jurisdiction) should also be located at the fire command station.

Automatic response is not required where the fire command station or a remote monitoring location is constantly attended by trained operators, and operator acknowledgement of receipt of a fire alarm signal is received within three minutes.

Emergency messages should be provided in the language of the predominant building population. Where there is a probability of isolated groups that do not speak the predominant language, multilingual messages should be provided.

Multichannel Capability

The system should allow the application of an evacuation signal to one or more zones and simultaneously permit voice paging to the other zones selectively or in any combination.

Functional Sequence

After an alarm signal is initiated, the system should automatically transmit:

1. An alert tone of 3 to 10 seconds duration followed by a message (or messages when multichannel capability is provided). The message should be given at least three times and direct the occupants of the fire initiation zone and other zones in accordance with the building's fire evacuation plan, or
2. An evacuation signal in accordance with the building's fire evacuation plan.

These signals should be transmitted immediately, or after a delay allowed by the authority having jurisdiction. Failure of the message above should sound the evacuation signal automatically. Provisions for manual initiation of voice instructions or evacuation signal generation should also be provided. Live voice instructions should override any previously initiated signals.

Voice and Tone Devices

The alert tone preceding any message can be part of the voice message, or transmitted automatically from a separate tone generator. When audio amplifiers and signal generators are provided, their output should be continuously monitored so an audible trouble signal results on failure.

The generating and amplifying equipment enclosed as integral parts and serving only a single listed loudspeaker need not be monitored. It is recommended to use backup amplifying and evacuation signal generating equipment with an automatic transfer feature so, upon primary equipment failure, prompt restoration of service in the event of equipment failure is ensured.

Voice/Alarm Communication System Components

Components of an emergency voice/alarm communication system typically include:

• Main Fire Alarm Panel: This system panel controls more functions, operations, inputs, and outputs than a traditional fire alarm panel. It typically has the following features:

1. An automatic alert tone with a preprogrammed sequence that can be transmitted throughout the building or to any selected zone.
2. A prerecorded evacuation message that is transmitted to the designated evacuation zone.
3. Two independent voice communication channels to operate the paging and alert tone systems. These channels include one-way voice communication capability to all areas ("all call") or zoned operation.
4. A terminal for the two-way fire department telephone.

• Primary Annunicator Panel: The annunciator, which is usually integrated with the control panel, can be a series of designated incandescent lights or LEDs, a graphic annunciator showing a building diagram, a CRT with a building diagram, or a video display such as a slide projector system. At a minimum, the annunciator panel should show the floor, zone, and type of alarm.

If the fire command station is integrated with other building systems (e.g., security, building controls) and is remote from the fire department response point, a separate annunciator panel for the voice communication system should be provided at the primary fire department response point. Additional annunciators may be necessary in larger buildings and should be located at secondary fire department response points.

• Remote Annunciators: These units generally do not have the same degree of detail as the primary annunciator panel and are provided as needed based on building size.

• Fire Department Two-Way Telephone Systems: The terminal for this system is located at the fire alarm command station. Remote stations can be jacks into which portable handsets are inserted or fixed (recessed or flush-mounted telephone stations).

• Remote Sounding Appliances: These consist of flush- or surface-mounted speakers, combination speaker/flashing emergency lighting unit, or reentry horns. The fire alarm alert tone and voice messages are generated from the fire alarm control panel.

• Input Devices: Manual fire alarm boxes, smoke and heat detectors, waterflow switches, and supervisory devices are typical input devices.

• *Peripheral Equipment and Controls:* These include elevator control, smoke control, and door unlocking control systems that usually incorporate annunciators, override switches, and status indicators.

Voice Communication Messages and Signals

There is no standardization of emergency voice/alarm communication messages or signals. Different vendors offer different types of signals, although there is not a substantial variation in the "temporal" and "slow whoop" alert tones and evacuation messages.

A voice message following an alert tone might be based on the following:

> "May I have your attention please. The building manager has
> directed all elevators to the entrance lobby. There has been a
> fire reported in the building, and the elevators may be needed.
> Please proceed to the lobby area for further instruction."

The message tells the occupants what is happening to the elevator, why this is happening, and what they are to do next. Each voice message, whether live or prerecorded, should be clear, calm, and informative and presented so it will motivate people into action. Voice messages should differ for certain parts of the building and certain stages of occupant evacuation.

Messages to the affected areas should follow a similar pattern. When a fire is reported on any floor, several other messages need to be transmitted once the elevator message has been sent. Most importantly, the occupants of the fire floor need to be told of the situation and instructed on actions to take. The adjacent floors need to be cleared, and these occupants must also be given instructions. Finally, a message must go to the "receiving" floors, to which these occupants will be sent.

Loudspeakers

Loudspeakers and their enclosures used in emergency voice/alarm communication signaling service should be listed for such service.

One or more loudspeakers should be located on each floor of the building and placed so their signals will be heard clearly, regardless of the maximum noise level from machinery or other equipment under normal occupancy conditions. Loudspeakers should also be provided for each elevator car and enclosed stairway in the building, and the speakers connected to separate paging zones.

Fire Command Station

A station should be located near a building entrance or other suitable location. The fire command station should provide a communication center for the arriving fire department and provide for control and display of the status of detection, alarm, and communications systems. The command station may be physically combined with other building operations and security centers as permitted by the local authority. Operating controls for fire department use should be clearly marked. A fire command station should control the emergency voice/alarm communication signaling service and the two-way telephone communication service where provided.

Monitoring Integrity of Conductors and Survivability

Because of its life safety applications, an emergency voice/alarm communication system must be in good working order at all times. Certain system conductors should be monitored, and the system designed for survivability in a fire emergency.

The following interconnecting conductors should be monitored: all means of interconnecting equipment, devices, and appliances, including standby battery and two-way telephone communication service equipment when provided. The occurrence of any single ground or open in the installation conductors should be automatically signaled to and received by the building fire command station as a trouble signal within 200 seconds.

An audible trouble signal should sound when a short or open in loudspeaker installation wiring occurs.

For an emergency system and where a two-way telephone communication circuit is provided, installation wires should be monitored for a short-circuit fault that could render the telephone communication circuit inoperative.

"Survivability" is defined as a system designed and installed so that attack by fire in a paging zone (causing loss of communication to this paging zone) should not result in loss of communication to any other paging zone. The system also should be so designed and installed that attack by fire causing failure of equipment or a fault on one or more installation wiring conductors of one communication path should not result in total loss of communication to any paging zone. Exceptions to survivability can be found in NFPA 72, *National Fire Alarm Code.*

Combination Systems

For combination systems that use common wiring, or equipment for combination systems, equipment other than that for emergency voice/alarm communication systems and fire protective signaling systems should be connected to the common wiring of the systems so that short, or open circuits, or faults in this equipment or between this equipment and the common wiring do not interfere with (1) the supervision of the fire systems or (2) prevent alarm or supervisory signal transmission for the fire systems.

In combination systems, a fire alarm signal should be clearly recognizable and take precedence over any other signal, even though a non-fire alarm signal is initiated first. Distinctive alarm signals should be obtained between fire alarm and other functions.

Two-Way Telephone Communication Service

A two-way telephone communication service is often provided in buildings for use in emergencies. When provided, two-way telephone communication service should be available for use by the fire service.

The service can also be used for signaling and communication for a building fire warden organization, signaling and communication for reporting a fire and other emergencies (e.g., voice call box service), signaling and communication for guard service, and other uses. Any variations in equipment and systems operation that create additional use of the two-way telephone communication service should not adversely affect its performance when used by the fire service.

Two-way telephone communication service should be capable of permitting simultaneous operation of any five telephone stations in a common talk mode.

Off-Hook Indicator: There should be a distinctive notification signal that indicates the off-hook condition of a calling telephone circuit at the fire command station. When a selective talk telephone system is supplied, there should be a distinct visible indicator for each selectable circuit so all off-hook conditions are continuously and visibly indicated.

Fire Service Use: There should be at the minimum a common talk circuit, i.e., conference or party line, for fire service use. The minimum requirement for fire warden use, where provided, should be a selective talking system controlled at the fire command station. Either system should be capable of operating with five telephone stations connected together. There should be a minimum of one fire service telephone station or jack per floor and a minimum of one per exit stairway and, where provided, at least one fire warden station or jack to serve each fire paging zone.

EMERGENCY CONTROLS

Emergency control or actuation of other systems in the building, e.g., air-handling systems, extinguishing systems, etc., is another form of signal notification. This function is usually required either by the local code and/or local authority.

NFPA *101, Life Safety Code,* sets certain requirements for emergency controls. A signaling system should be arranged to automatically actuate control functions necessary to make the protected premises safer for building occupants.

The fire alarm system can actuate the following emergency controls:

1. Elevator capture and control, usually in accordance with ASME/ANSI A17.1, *Safety Code for Elevators and Escalators;*
2. Release of automatic door closers;
3. Stairwell or elevator shaft pressurization;
4. Smoke management or smoke control systems;
5. Initiation of automatic fire extinguishing equipment;
6. Emergency lighting control;
7. Unlocking of doors; and
8. Emergency shut off for gas and fuel supplies that may be hazardous, provided continuation of service is not essential to life safety.

Performance of emergency control functions must not impair the effective response of all required alarm notification functions. It is recommended that wiring to required emergency control devices be electrically supervised to within 3 ft (0.9 m) of the device to be actuated.

Location

Operator controls, visual alarm annunciators, and manual communications capability for the emergency control function should be installed in a control center at a convenient location. Controls used solely by the fire department should be located adjacent to an entrance at a fire command station or as designated by the authority having jurisdiction.

Automatic detection should be provided for the central controls when controls are not in continuously occupied areas. A detector should be provided for each central control location.

Elevator Recall

Any smoke detectors used in elevator lobbies and to recall elevators for fire fighter's service should be connected to the building fire alarm system.

An elevator lobby smoke detector should be capable of initiating an alarm when all other devices on the same alarm initiating circuit have been manually or automatically placed in the alarm condition. (Elevator recall can also be provided by the fire alarm control equipment.)

When actuated, each elevator lobby smoke detector or smoke zone should initiate an alarm condition on the building fire alarm system. The alarm initiation circuit or zone from which the alarm originated should be indicated at the control panel and required remote annunciators. Elevator lobby smoke detectors, as with all automatic fire detectors, should be tested regularly.

Smoke detectors for elevator recall should be installed in the elevator lobbies and elevator machine room. Even if facilities are not equipped with a fire alarm signaling system, elevator recalls are required to be controlled with a smoke detection system.

ANSI-A17.1, *Safety Code for Elevators and Escalators*, which is referenced by NFPA *101*, *Life Safety Code*, states that smoke detectors for each group of elevators serviced by the same call button be provided with two elevator detection circuits. These circuits extend to the elevator machine room, where the elevator controller is located. Typically, the circuits use normally closed dry contacts. In the event of an electrical failure of the circuit, the elevator recall is activated upon a loss of signal.

Detection circuits should be required in lobbies, hoistways, and machine rooms. Upon activation of one of the detectors on this circuit, the elevator responds to a designated floor that is specified by the authority having jurisdiction to be a safe egress level. An additional smoke detection circuit covers the elevator lobby designated recall level. When this level smoke detector activates, the elevator responds to an alternative recall level, as determined by the authority having jurisdiction. Usually, the hoistway smoke detectors are also connected to this second detection circuit.

Activated smoke detectors recall all elevators to the designated recall level, opening all elevator doors, while all call buttons become inoperative. These elevators then become available to the fire department personnel for evacuation and fire fighting functions.

Release of Automatic Door Closers

A door designed to be kept normally closed in a means of egress, such as a door to a stair enclosure or horizontal exit, should be a self-closing door and should not at any time be secured in the open position.

Circumstances where alternatives to self-closing doors can be used are found in NFPA *101*, *Life Safety Code*, and are permitted by this Code for (1) buildings with low- or ordinary-hazard contents or (2) when permitted by local authorities. Doors can be released by an automatic smoke detection system; an automatic sprinkler system can activate the door closing, but does not substitute for smoke detector actuation.

Activation of Mechanical Ventilation Systems

Both mechanical ventilation and pressurized stair enclosure systems can be initiated by a smoke detector installed in an approved location within 10 ft (3 m) of the entrance to the smokeproof enclosure. The system can also be activated by manual controls accessible to the fire department and waterflow signal from a complete automatic sprinkler system or a general evacuation alarm signal. Activation of the closing device in any door should activate the closing devices on all doors in the smokeproof enclosure at all levels.

Smoke Management

Each air-transfer opening or duct penetration of a required smoke barrier should have an approved damper designed to resist the passage of smoke. NFPA *101*, *Life Safety Code*, permits exceptions to provisions for smoke dampers; they may be omitted:

1. In ducts or air-transfer openings that are part of an engineered smoke control system;
2. In ducts where the air continues to move and the air-handling system installed prevents recirculation of exhaust or return air under fire emergency conditions; or
3. Where ducts penetrate floors that serve as smoke barriers.

NFPA *101*, *Life Safety Code*, also allows smoke dampers to be omitted in other circumstances.

Smoke dampers in ducts penetrating smoke barriers should close upon detection of smoke by the facility's smoke detectors. They should also close on detection of smoke by local smoke detectors on either side of the smoke barrier door opening (when ducts penetrate smoke barriers above the smoke barrier doors), or smoke detectors located within the ducts in existing installations. For details on smoke management, refer to NFPA 72, *National Fire Alarm Code*; NFPA 90A, *Standard for the Installation of Air Conditioning and Ventilating Systems*; and NFPA 101, *Life Safety Code.*

Automatic Shutdown

In systems of more than 2,000 cu ft/min (944 L/sec) capacity, listed smoke detectors should be installed to detect the presence of smoke. The detectors should automatically stop the fan(s) in the supply system downstream of the filters, in the return system on each floor at the point of entry into the common return, or in a system of smoke detectors providing total area coverage. A smoke detector in the return air stream is not essential in systems of less than 15,000 cu ft/min (7080 L/sec) capacity.

Detectors in the supply and return air stream may also be omitted, provided that the system is (1) less than 15,000 cu ft/min (7080 L/sec) capacity, (2) the entire system is within the space served, and (3) such space is protected by an area smoke detection system.

Smoke Dampers

Smoke dampers should be installed in systems over 15,000 cu ft/min (7080 L/sec) capacity to isolate the air handling equipment (including filters) from the remainder of the system. This restricts circulation of smoke and is arranged to close automatically when the system is not operating.

The protection provided by the installation of smoke detectors and automatic shutdown is intended to prevent the dispersion of smoke through the supply air duct system and, preferably, to exhaust a significant quantity of smoke to the outside. However, these functions will not guarantee early detection of fire or the detection of smoke concentrations prior to dangerous smoke conditions if smoke moves other than through the supply air system. Where facility smoke control protection is needed, a specifically designed smoke control system should be provided that includes placement of smoke detectors so that early detection of developing smoke conditions and initiation of the smoke control measures are provided.

Ideally, actuation of any smoke detector should sound the fire alarm and provide for control of the ventilation systems. NFPA 72, *National Fire Alarm Code*, provides requirements for installation and connections. Activation of air duct smoke detectors installed in a building that is not equipped with a building fire alarm system should also sound an audible alarm in a normally occupied area. A trouble condition should be indicated visibly or audibly in a normally occupied area and identified as air duct detector trouble.

Automatic Fire Extinguishing Equipment

The potential fire loss of a facility may require a certain type of automatic extinguishing system in addition to an automatic sprinkler system. Examples of automatic extinguishing systems are carbon dioxide, dry chemical, foam, Halon 1301, water spray, or a standard extinguishing system of other type.

Emergency Lighting

Emergency lighting should be provided in the stair shaft and vestibule. A standby generator installed for the smokeproof enclosure mechanical ventilation equipment may be used as the power supply.

Unlocking of Doors

Stairwell doors should allow reentry from the stairwell to the interior of the building, or an automatic release provided to unlock all stairwell doors. Such automatic release should be actuated with the initiation of the building fire alarm system.

Special Locking Arrangements

For buildings with a supervised automatic fire detection system or sprinkler system, doors in low- and ordinary-hazard areas can be equipped with approved, listed, locking devices. The devices should:

1. Unlock upon actuation of a supervised automatic fire detection system or sprinkler system, and
2. Unlock upon loss of power controlling the lock or locking mechanism, and
3. Release the lock within 15 seconds whenever a maximum force of 15 lb (67 N) is continuously applied to the release device for not more than three seconds. Such doors should be relocked manually. An operation signal in the vicinity of the door should activate when the lock releases.

A sign on the door adjacent to the release device should be provided, stating:

PUSH UNTIL ALARM SOUNDS. DOOR CAN BE OPENED IN 15 SECONDS.

Emergency lighting should also be provided at the door.

Emergency Shut Off

Emergency shut off of fuel supplies is usually achieved with a solenoid-operated valve.

SIGNAL NOTIFICATION IN HIGH-RISE BUILDINGS

Delayed detection of fire is a significant problem in high-rise buildings. Early discovery of fire is important to ensure prompt evacuation, but total evacuation within reasonable time limits is not always possible in high-rise buildings. Exits are not designed to handle the total occupant load of a high-rise building at one time. A 44-in. (1.12-m) stairway will discharge about 140 persons per minute, so only 140 persons can enter the stairway per minute once the stairway has been filled with people. Most building codes do not require the width of exits to increase during descent toward the stair discharge. In analyzing egress in high-rise buildings, it becomes obvious that the total evacuation of buildings over a certain height [generally 150 ft (45 m) or 12 stories] becomes unreasonable.

Further, while occupants may be trying to evacuate the building, fire fighters may need to use the stairs at the same time to reach the fire floor. For this and other reasons, partial evacuation is usually a more realistic approach than complete evacuation in high-rise buildings.

There are two options that can be used to alert occupants of a fire in a high-rise building: (1) total and (2) selective. Total building alerting capability should be provided if it is desired to alert all occupants. For selective evacuation, the fire alarm system designer and code officials should decide the number of occupants to be alerted and the method to be used. The fire area should also be identified in some manner, since fire fighters cannot effectively search an entire building for the fire site. The fire department and building management should be notified automatically when the alarm system is activated.

Emergency voice/alarm communciation system design in a high-rise occupancy is based on occupant load, occupant physical characteristics, active/passive protection systems, manual fire fighting capability, and fire department access. Any voice/alarm system for a high-rise building should be capable of:

1. Notifying the fire command station of the existence and location of the fire;
2. Notifying the fire department of the existence of a fire;

3. Directing occupants of the fire zone to the outside or to an area of refuge;
4. Delaying transmission of alarms outside the immediate fire area until an authority in charge considers it appropriate to sound alarms to other areas;
5. Operating in a simple and reliable manner;
6. Monitoring essential interconnecting equipment and components of the system; and
7. Allowing a paging zone to survive the fire loss of another paging zone.

The audible fire alarm signal to the occupants of the fire zone can include continuous sounding devices, voice direction that momentarily silences the continuous sounding devices, or both.

A message is typically sent to the fire area and floors above and below the fire area. Facilities should be provided at the fire command station so voice and sounding appliance alarms can be manually activated in any combination of zones. The voice system should be capable of transmitting to all zones simultaneously; toward this, the emergency voice system should not be part of the same system as the building music/paging systems. Regardless of other signals or sounds, e.g., music, announcements, etc., on the system, the fire calls should automatically override all other system traffic and sound at maximum volume. The volume of a fire call should comfortably exceed the volume of any other signal or system sounds.

If the fire command station is integrated with other building systems, e.g., security and building controls, and is remote from the fire department response point, a separate annunciator panel for the voice/communication system should be provided at the primary fire department response point. Additional annunciators may be necessary in larger buildings at secondary fire department response points.

BIBLIOGRAPHY

NFPA Codes, Standards, Recommended Practices, and Manuals. (See the latest *NFPA Catalog* for availability of current editions of the following documents.)
NFPA 72, *National Fire Alarm Code.*
NFPA 90A, *Standard for the Installation of Air Conditioning and Ventilating Systems.*
NFPA *101, Life Safety Code.*

Additional Reading

ANSI A17.1, *Safety Code for Elevators and Escalators*, American National Standards Institute, New York, NY.

ASME/ANSI A17.1, *Safety Code for Elevators and Escalators*, American Society for Mechanical Engineers, New York, NY.

Butler, H., Bowyer, A., and Kew, J., *Locating Fire Alarm Sounders for Audibility*, Building Services Research and Information Association, Bracknell, 1981.

Schifiliti, R. P., Chapter 3-1, "Design of Detection Systems," *The SFPE Handbook of Fire Protection Engineering*, 1st ed., NFPA, Quincy, MA, 1988.

8

Power Supplies

INTRODUCTION

Certain conditions and monitoring provisions are important to ensure that power supplies adequately support a fire alarm signaling system. This chapter discusses factors to be considered in specifying sources for primary (main), secondary (standby), and trouble power supplies; identifying types of storage batteries and proper testing and maintenance procedures; and calculating adequate battery size for a fire alarm signaling system. Additional information can be found in Chapter 6, "Electrical Supervision," and Chapter 14, "Testing."

The general power supply requirements for a fire alarm signaling system according to NFPA 72, *National Fire Alarm Code*, are summarized in Chapter 2. All power supplies should be installed in conformance with requirements of NFPA 70, *National Electrical Code*, for such equipment.

Fire alarm signaling systems have three basic sources of power to operate the system: (1) a primary (main) supply to provide normal operating power; (2) a secondary (standby) power supply that must supply power when the main supply fails; and (3) a power supply to activate a trouble signal, which is required by NFPA 72, *National Fire Alarm Code*. The trouble power supply source is often the secondary supply.

A trouble signal, while required, does not need to indicate that the primary (main) power is being supplied by either of the two secondary (standby) sources of power as long as periods of operation are provided as specified in the secondary (standby) requirements of Table 2-2 and loss of primary power is otherwise indicated (e.g., by loss of building lighting).

PRIMARY (MAIN) POWER SUPPLY

The primary (main) power supply should have a high degree of reliability, and consist of a light and power service or an engine-driven generator. The primary (main) power supply should be capable of operating the system continuously for the maximum connected load with all initiating devices and alarm indicating appliances in an alarm state.

Light and Power Service

A light and power service that operates the system under normal conditions should have a high degree of reliability and capacity for the intended service. Connections to the light and power service should be on a dedicated branch circuit that is mechanically protected with circuit disconnecting means accessible only to authorized personnel and clearly marked "fire alarm circuit control."

Engine-Driven Generator

An engine-driven generator should be used only where a person specifically trained in its operation is on duty at all times. Installations of engine-driven generators are covered by provisions of NFPA 110, *Standard for Emergency and Standby Power Systems*. A 24-hour operator is not required for multiple generators or a single generator with storage battery capacity.

When gasoline is used as main fuel for the generator, it should be stored in outside underground tanks whenever possible. Gravity tanks should not be used. Sufficient fuel for generator use should be available in storage for six months of testing plus 24 hours of operation at full load. If additional fuel can be obtained on two hours notice, fuel storage can be cut to 12 hours of operation at full load. If natural or manufactured gas via reliable utility mains is used as generator fuel, fuel storage tanks are optional, unless gas is piped through or to an earthquake-prone area.

A separate storage battery and separate automatic charger should be provided for starting the engine-driven generator and not used for any other purpose.

Primary Power in Household Fire Warning Equipment

Power supplies for this equipment should be sufficient to operate the alarm signal(s) for at least four continuous minutes.

An ac primary source of electric power, if used, should be a dependable commercial light and power supply source. A visible "power on" indicator should be provided.

If a qualified electrician does not install the household system, it is advisable that the power source not exceed 30 volts. Requirements for these circuits can be found in NFPA 70, *National Electrical Code*, Article 760.

A secondary source of power for household equipment is desirable. When a secondary source is provided, it should be able to operate the system for 24 hours and thereafter to sound alarm devices for not less than four minutes.

A restraining means should be used at the plug-in of any cord-connected installation. Household fire warning equipment should not be subject to loss of power by a wall switch. Neither loss nor restoration of primary power should cause an alarm signal.

Single-station and multiple-station smoke detectors that are powered from 120-volt ac sources should not be installed on circuits protected by a ground-fault circuit-interrupter.

When a standby battery is provided, the primary power supply should be of sufficient capacity to operate the system under all conditions of loading with any standby rechargeable battery fully discharged or nonrechargeable-type battery disconnected.

SECONDARY (STANDBY) POWER SUPPLY

The secondary (standby) power supply should supply the energy to the system automatically within 30 seconds whenever the primary (main) power supply is incapable of providing the minimum voltage required for proper operation. The secondary (standby) power supply should not activate as long as the primary power supply voltage remains above 85 percent of rated voltage. The secondary (standby) power supply should be capable of operating the system under maximum normal load for the specified standby time and subsequently for five minutes in the alarm condition. The secondary (standby) power supply should consist of:

1. A storage battery and charger; or
2. An engine-driven generator and storage batteries with four hours capacity; or
3. Multiple engine-driven generators, one of which is arranged for automatic starting. The multiple generators should be capable of supplying the energy required by the secondary power supply with the largest generator out of service. (The second generator may be pushbutton start.)

OTHER POWER SUPPLIES

A separate power supply, independent of the primary (main) power supply, should be provided to operate trouble signals when the primary (main) power supply fails. A primary battery (dry cell) is not permitted to power the trouble signals. The secondary (standby) power supply can be used for this purpose. Additional power supplies, when provided for control units, circuit interfaces, or other equipment essential to system operation, that are located remote from the main control unit should also have the same power supplies (main/secondary/trouble) as provided for the main control equipment.

Rectifiers

A rectifier used in a power supply circuit without a floating battery should be approved for the purpose and of adequate capacity to maintain voltage regulation between 130 percent of rated voltage at no load and 100 percent of rated voltage at maximum load.

STORAGE BATTERIES

The battery should be fully charged for all conditions of normal operation and automatically maintained in a charged condition. After the fully charged battery is subjected to a single discharge cycle, there should be sufficient charging capacity to restore the battery within 48 hours. The trickle charge rate should not be excessive, which can cause battery damage after a fully charged condition is achieved.

Storage batteries should be located or enclosed so the equipment of the signaling system (including overcurrent protective devices) will not be affected adversely by battery gases. Installation of the batteries is further described in NFPA 70, *National Electrical Code*, Article 480.

The method used to charge a battery should provide either integral meters or readily accessible terminal facilities. This allows for the connection of portable meters to determine the battery voltage and charging current. The storage battery should be protected by overcurrent devices with a 150 percent minimum/200 percent maximum rating of the battery operating load.

Periodic tests should be performed to ensure the secondary supply batteries are in good working condition at all times. Performance criteria should be used to test nonsealed lead-acid and nickel-cadmium batteries. Tables 8-1 and 8-2 test methods are derived from NFPA 1221, *Standard for the Installation, Maintenance, and Use of Public Fire Service Communication Systems*, and are applicable to any fire alarm signaling system that uses these types of batteries as a standby power supply.

TABLE 8-1. Test Intervals for Lead-Acid Batteries

	Maximum Test Interval
Measure Float Voltage	
Of entire battery or a pilot cell	1 week
Of each cell	3 months
Measure Specific Gravity	
Of a pilot cell	6 weeks
Of each cell	6 months
Discharge for 2 hrs	1 year
Clean and Inspect	3 months
Calibrate Meters	1 year

To maximize battery life, the battery voltage for lead-acid cells should be maintained within the limits shown below:

Float Voltage	High-Gravity Battery (Lead-Calcium)	Low-Gravity Battery (Lead-Antimony)
Maximum	2.25 volts/cell	2.17 volts/cell
Minimum	2.20 volts/cell	2.13 volts/cell

NOTE: High-rate voltage is 2.33 volts/cell. Both high- and low-gravity voltage is (+)0.07 volts and (-)0.03 volts.

TABLE 8-2. Test Intervals for Nickel-Cadmium Batteries

	Maximum Test Interval
Measure Float Voltage	
(1.42 volts per cell, nominal)	
Of entire battery	3 months
Of each cell	1 year
Check State of Charge	6 months
Discharge for 2 hrs	1 year
Clean and Inspect	3 months
Calibrate Meters	1 year

To maximize battery life, the battery should be charged as follows:

Float Voltage	1.42 volts/cell ±0.01 volts
High-Rate Voltage	1.58 volts/cell +0.07–0.00 volts

Additional factors should be considered when checking the state of charge for nickel-cadmium batteries. If the battery charger is switched from float to high-rate mode, the current (as indicated on the charger ammeter) will immediately rise to the maximum output of the charger. The battery voltage as shown on the charger voltmeter will start to rise at the same time. Since many variables are involved, the actual value of the voltage rise is unimportant. The length of time it takes for the voltage to rise is the important factor.

For example: if the voltage rises rapidly in a few minutes, and holds steady at the new value, the battery is fully charged. At the same time, the current will drop to slightly above its original value. In contrast, if the voltage rises slowly and the output current remains high, the high-rate charge should be continued until the voltage remains constant. This condition indicates that the battery was not fully charged, so the float voltage should be increased slightly.

Rechargeable (Storage) Type Batteries

Three types of rechargeable batteries are normally used in fire alarm signaling systems: (1) vented lead-acid gelled or starved electrolyte, (2) nickel-cadmium, or (3) sealed lead-acid.

A vented lead-acid gelled or starved electrolyte battery is generally used instead of primary batteries in applications with a relatively high current drain (up to 1 ampere), or that require extended standby capability of much lower supervisory currents. Nominal voltage of a single cell is 2 volts, and the battery is available in multiples of 2 volts (2, 4, 6, 12, etc.). These batteries should be fully charged before storage, stored in an upright position, and stored in areas with maximum temperatures of 100°F (37.8°C). Additionally, batteries stored at temperatures of 70°F (21°C) or lower should be recharged once per year. At higher temperatures, more frequent recharge is required. If improperly stored, the batteries will not achieve full capacity or be rechargeable.

The battery should be installed in a clean, dry place accessible for servicing at a maximum ambient temperature of 70°F (21°C). Battery life is cut in half for each 20°F (11.1°C) increase in continuous ambient temperature.

This rechargeable-type battery is usually maintained in a fully charged condition by a continuous float-charge-type rectifier.

A nickel-cadmium battery (sealed type) is generally used where the battery current drain during a power outage is low to moderate (up to a few hundred milliamperes), and fairly constant. Nickel-cadmium batteries are also available in much larger capacities. The normal voltage per cell is 1.42 volts, and the battery is available in multiples of 1.42 volts (12.78, 25.56, etc.).

These batteries can be stored in any state of charge for an indefinite period. However, they will lose capacity "self-discharge" depending on storage time and temperature. Typically, batteries stored for more than one month will require an 8- to 14-hour charge period to restore capacity. While in service, the battery should receive a continuous constant charging current sufficient to keep it fully charged. (Typically, the charge rate equals $\frac{1}{10}$ to $\frac{1}{20}$ of the ampere-hour rating of the battery.) Because the batteries are composed of individual cells connected in series, it is possible, during deep discharge, that one or more low-capacity cells may be completely discharged before other cells. When this happens, the cells with remaining life tend to "charge" the depleted cells, causing a polarity reversal that results in permanent battery damage. This reversed polarity condition can be determined by measuring the open-cell voltage of a fully charged battery. Voltage should be a minimum of 1.28 volts per cell multiplied by the number of cells.

Voltage depression effect is a minor change in discharge voltage level caused by constant current charging below the system discharge rate. For example: if a shaving device with nickel-cadmium batteries is used repeatedly for one minute per day; followed by a recharge, the ampere-hour output cannot be achieved because the battery, in this instance, has developed a one-minute memory.

With a sealed lead-acid battery, the electrolyte is totally absorbed by the separators and normally no venting occurs. Gas evolved during recharge is internally recombined, resulting in minimal loss of capacity life. A high-pressure vent is provided to avoid damage under abnormal conditions.

Other battery characteristics are comparable to the vented lead-acid battery described above.

At temperatures below 70°F (21°C), less than rated capacity will be available depending on discharge rate. Manufacturer's data should be consulted in these situations.

Calculating Battery Size

The required battery capacity is determined by the work it is expected to perform and expressed in ampere-hours [i.e., the number of amperes required, multiplied by the time (in hours) the battery is to be used].

Two factors should be considered in calculating battery size:

1. The power required to operate the system in normal supervisory condition ("supervisory current in amperes") for a certain number of hours (ampere-hours) with the primary (main) power supply disconnected, and
2. After the time period in 1 (above), power in ampere-hours required to operate the system in alarm condition ("alarm current") for a specified time, usually 5 minutes.

Calculations of battery size usually involve determining and adding together dc current requirements.

To obtain supervisory current, determine and add together:

1. Normal supervisory current of the main control panel,
2. The supervisory current of each alarm initiating zone multiplied by the number of zones on the system,
3. The supervisory current of each alarm indicating zone multiplied by the number of system zones,
4. The current required to operate the "trouble silenced" feature, and
5. Any other supervisory current (e.g., fire house connection, shut down relays, etc.) normally required to operate the system.

Then, determine the total alarm current, which is the operating current of each alarm indicating appliance (bells, horns, etc.) multiplied by the number of appliances on the system.

For example: if the supervisory current of a fire alarm system is 0.3 amperes (300 ma) and must be supplied for a 24-hr period, 7.2 ampere hours (AH) would be needed ($0.3 \times 24 = 7.2$ AH). Then, if the alarm current is 2 amperes for 5 minutes ($\frac{1}{12}$ hr), the needed capacity would be 0.17 AH ($2 \times 0.085 = 0.17$). The total battery size needed is 7.2 AH for 24 hours plus 0.17 AH for 5 minutes for 7.37 AH. An 8-AH battery would fit the need for this particular system.

To ensure selection of the proper size battery, the battery manufacturer's discharge curves should be consulted. Battery manufacturers usually rate each of their batteries in ampere hours at a specific rate, such as 8 AH or 20 AH.

Figure 8-1 shows the discharge characteristics of a lead-acid battery. Note that the ampere hour rate of a battery changes with the applied load and the amount of time it is applied.

TECHNICAL DATA, OVERALL DIMENSIONS & WEIGHTS
PREFIX NUMBER INDICATES CELLS PER UNIT . . . . . .SUFFIX NUMBER DENOTES PLATES PER CELL

UNIT TYPE	CELLS PER UNIT	CAPACITIES								OVERALL DIMENSIONS IN INCHES PER UNIT			APPROXIMATE WEIGHT PER UNIT IN POUNDS		ELEC-TROLYTE GALLONS PER CELL
		USUAL RATES					SPECIAL RATES								
		IN AMPERE HOURS			IN AMPERES		IN AMPERE HOURS								
	(4 UNITS REQ'D)	8 HR RATE TO 1.75 FV	8 HR RATE TO 1.75 FV	1 HR RATE TO 1.75 FV	1 MIN RATE TO 1.75 FV	1 MIN RATE TO 1.50 FV	72 HR RATE TO 1.85 FV	8 HR RATE TO 1.92 FV	100 HR RATE TO 1.75 FV	LENGTH	WIDTH	HEIGHT	NET	PACKED	
3CPE3	3	8.0	6.2	4.0	23	37	10.8	5.7	15	6-9/16	4-5/16	8-1/8	12	17	0.130
3CPE5	3	16.0	12.3	8.0	46	74	21.6	11.5	30	6-9/16	4-5/16	8-1/8	15	22	0.115
3CPE7	3	24.0	18.5	12.0	69	110	32.4	17.2	—	8-3/8	4-3/16	8-3/4	22	26	0.161

ALL RATINGS INCLUDE VOLTAGE DROP ACROSS INTER-CELL CONNECTORS USED IN STANDARD LAYOUTS.
FV = FINAL VOLTAGE.
FOR S1 UNITS = 1 IN. = 25.4 MM; 1 LB = 0.45 KG; 1 GAL = 3.9 L.

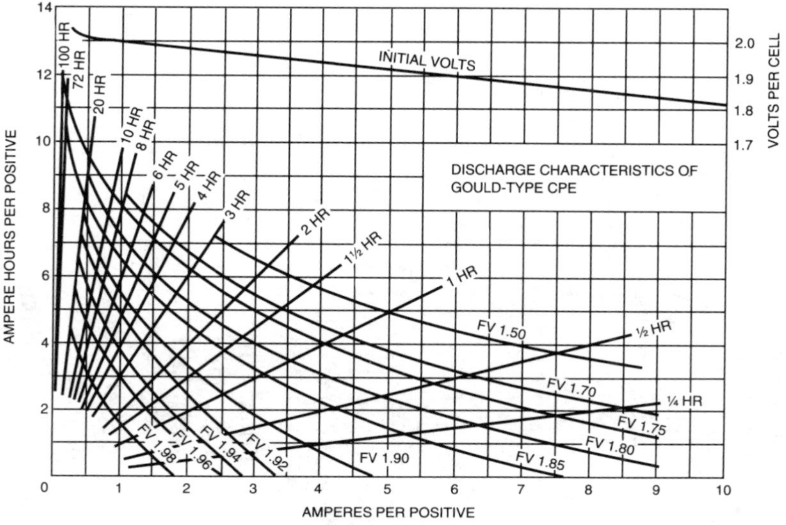

Figure 8-1. Discharge characteristics of a lead-acid battery.

A fully charged battery can provide a certain current, in amperes, for a specific period of time. Figure 8-1 indicates that the 3CPE3 battery, at the 8-hour rate, is an 8-AH battery. This means that it will supply a current of 1 ampere for 8 hours and at the 8-hour mark each cell will have a final voltage (FV) of approximately 1.75 volts. However, if the current drain is doubled to 2 amps, the battery will be reduced to 1.75 volts in approximately 3 hours. Effectively, at this discharge rate, the battery is able to deliver only a little more than 6 AH.

In the special rates in Figure 8-1, using the same battery, a discharge rate of 0.15 amps for 72 hours would raise the AH capacity of the battery to 10.8 AH.

This change in AH capacity of a battery must be considered when sizing a battery for a specific system. To compensate for decreased capacity with aging batteries, it is advisable to select a larger battery than required by calculations especially if the calculated AH is equal to or just below the battery AH capacity.

PRIMARY (DRY CELL) BATTERIES

According to NFPA 72, *National Fire Alarm Code*, primary (dry cell) batteries are permitted to be used for specific service. In central station signaling systems, NFPA 72 allows a primary battery to be used for the trouble signal power supply as follows:

1. Primary batteries should be located in a clean dry place that is accessible for servicing and where the ambient air temperature is 40°F (4.4°C) minimum and 100°F (37.8°C) maximum.
2. The maximum normal load for a No. 6 primary battery should not exceed 2 amperes per cell. No. 6 batteries should be replaced under the following conditions:

(a) An individual primary battery cell rated 1.5 volts should be replaced when a test load of 1 ohm reduces the potential below 1 volt.

(b) A unit assembly of primary battery cells rated 6 volts should be replaced when a test load of 4 ohms reduces the potential of the unit below 4 volts.

Battery Use in Low-Power Wireless Initiating Device Transmitters

Primary batteries can also be used as the sole power source in low-power wireless initiating device transmitters. When a primary battery is used in this manner, each transmitter should serve only one device and be individually identified at the receiver/control unit. The battery should be capable of operating the wireless initiating device transmitter for a minimum of one year before the battery depletion threshold is reached.

A battery depletion signal should be transmitted before the battery reaches a level that cannot support alarm transmission after seven additional days of normal operation. This signal should be distinctive from alarm, supervisory, tamper, and trouble signals; should visibly identify the affected initiating device transmitter; and, if silenced, automatically resound at least once every four hours.

Catastrophic (open or short) battery failure should cause a trouble signal that identifies the affected initiating device transmitter at its receiver/control unit. If silenced, the trouble signal should automatically resound at least once every four hours.

Failure of a primary battery in any way in one initiating device transmitter should not affect any other initiating device transmitter.

Battery Use in Household Fire Warning Equipment

Although ac power is usually required and preferred for household fire warning equipment in all new construction, primary batteries can be used under certain conditions. A battery used for household fire warning equipment should be monitored to ensure that the following conditions are met:

1. All power requirements should be met for at least one year's battery life (including monthly testing).
2. A distinctive audible trouble signal should sound before the battery is incapable (from aging, terminal corrosion, etc.) of operating the device(s) for alarm purposes.
3. Automatic transfer should be provided from alarm to a trouble condition for a unit employing a lock-in alarm feature.
4. The unit should be capable of producing an alarm signal for at least 4 minutes at the battery voltage at which a trouble signal is normally obtained, followed by seven days of trouble signal operation.
5. The audible trouble signal should be produced at least once every minute for seven consecutive days.
6. The monitored batteries meeting these specifications should be clearly identified on the unit near the battery compartment.

BIBLIOGRAPHY

NFPA Codes, Standards, Recommended Practices, and Manuals. (See the latest *NFPA Catalog* for availability of current editions of the following documents.)

NFPA 70, *National Electrical Code*.

NFPA 72, *National Fire Alarm Code*.

NFPA 110, *Standard for Emergency and Standby Power Systems*.

NFPA 1221, *Standard for the Installation, Maintenance, and Use of Public Fire Service Communication Systems*.

Additional Reading

UL 985, *Household Fire Warning System Units*, Underwriters Laboratories Inc., Northbrook, IL.

UL 1023, *Household Burglar-Alarm System Units*, Underwriters Laboratories Inc., Northbrook, IL.

9

Fire Detection System Design

INTRODUCTION

The importance of proper design for fire alarm signaling systems cannot be overemphasized. This chapter reviews design objectives for fire alarm and detection systems and covers criteria necessary to design a proper, effective system. Design criteria discussed in this chapter include consideration of combustibles, possible fire scenarios, and the effects of ceiling height and configuration, and room ventilation and temperature on system design.

Criteria for the selection and spacing of heat, smoke, and flame-sensing detectors for various construction types are addressed herein. Further information on signaling system design can be found in Chapter 13, "Installation."

Fire protection design requires implementation of both engineering design and administrative controls. Fire hazards are controlled through fire prevention, fire protection program maintenance, and training.

Fire prevention is the preferred method to control fire loss in a facility. Since it is unlikely that all fires can be prevented, administrative control of fire hazards, combined with engineering design controls, can greatly reduce both the frequency and severity of fire. Fire prevention is accomplished by controlling both combustibles and ignition sources.

Combustibles can be controlled by locating them remote from possible ignition sources, containing them within suitable enclosures, and selecting noncombustible materials. Ignition sources can be controlled by use of specialized electrical equipment for hazardous locations (those involving flammable gases, vapors, combustible dusts, and ignitable fibers), minimizing static electricity buildup potential by using grounding techniques and maintaining high humidity, and designing electrical systems according to NFPA 70, *National Electrical Code.*

An administrative control program can be implemented through operational and maintenance procedures and through the use of a permit system to control hazards.

Fire protection by design focuses on managing the impact of fire, since fire prevention cannot always be obtained. Two major categories of fire protection recognized today are: (1) passive fire protection and (2) active fire protection. Passive fire protection centers on the control of fire through building design. For a passive system, design objectives for building design include:

1. Separate hazardous areas from essential areas in the facility layout.
2. Compartmentalization with fire barriers to limit fire spread.
3. Use of fire-retardant materials to reduce fire growth and spread.

Passive fire protection should begin in the conceptual design stage and be carried on through design and construction. Retrofitting a facility with passive fire protection is normally a very expensive and difficult process.

Active protection involves fire control with specialized fire protection systems and equipment. The general design objective for active protection is the detection and suppression of fire before unacceptable damage can occur. Two key types of active fire protection are: (1) fire suppression systems and (2) fire detection systems. Common active fire suppression systems for the control and extinguishment of fire are automatic sprinkler systems, fixed water spray systems, halon systems, CO_2 systems, and foam extinguishing systems. Heat, smoke, and flame-sensing devices are examples of fire detection systems.

Most extinguishing systems have fire detection components that automatically actuate the system.

DESIGN OBJECTIVES

The design objectives for a fire alarm and/or a fire detection system must include the basic functions of the system. These functions can include notifying building occupants of emergency conditions for evacuation purposes; alerting organized assistance, such as a fire brigade or fire department, to undertake fire fighting operations; detecting specific stages, e.g., smoldering or flaming, of fire development; actuating fire suppression systems; supervising fire suppression systems; and supervising processes for abnormalities that might cause fire.

DESIGN CRITERIA

Design criteria provide the parameters for system specification and design drawings. Preparation of design criteria is essential to the fire alarm signaling system design package, and must be provided prior to the construction drawings and detailed specifications. The criteria should include an area-by-area engineering study that defines the type of combustibles in each area and postulates the likely fire scenarios. The construction features of the facility are of the utmost importance when considering the design of the fire alarm signaling system. Ceiling heights and the type of ceiling construction are two features that must be examined for optimum detector placement. Other building parameters that must be analyzed are the ventilation and temperature for each area. Natural and mechanical air flow patterns must be defined and quantified during the analysis state of design.

Design criteria address system function and overall operation. Again, the design criteria must be prepared prior to the design of a fire alarm signaling system, and include technical data defining system purpose and design objectives, design parameters, applicable design codes, standards and regulations, and quality assurance requirements.

The two essential parameters for fire detection are: (1) combustible types and (2) possible fire scenarios. These must be evaluated through an area-by-area analysis of a facility.

1. Types of combustibles in each area. Each area of the facility where fire detection may be used should be reviewed to determine the type of fuel present, i.e., solids, liquids, or gases. The type of fuel will often dictate the type of fire detection needed. For example, if the design objective is evacuation of the building occupants, and the combustible material in the area is smoke-producing material, the design would call for smoke detectors rather than heat detectors. If the design objective of the fire detection system in a garage facility is to actuate a deluge sprinkler system (because the garage may contain exhaust gases), the system design should consider specifying use of heat or photoelectric smoke detectors rather than ionization-type smoke detectors.

2. Types of possible fire scenarios. Fire scenarios should be based on the type of combustibles in the area. Fires can be categorized into smoldering or flaming types. Fire growth varies as a function of types of combustibles, their physical arrangement, surface-to-mass ratio, and ignition source energy. Some common types of postu-

lated fires include those involving ordinary combustibles, e.g., cellulosic products, flammable/combustible liquids; electrical cable insulation; plastic fuel; and combustible metals, dusts, or fibers. The types of likely fire scenarios need to be evaluated prior to selection of the actual detector.

Other design parameters that should be evaluated in the design criteria phase of a project include ceiling height, stratification, ceiling configuration, room ventilation, and room temperature.

Design Parameters

Ceiling Height: Since smoke, heat, and fire gas production in the early stages of fire is relatively small in most structural fires, the effect of ceiling height is significant when a fire begins. As ceiling height increases, a larger fire is needed to actuate the same detector in the same time. As fire increases in intensity and severity, the significance of the effect of ceiling height is reduced due to the buoyancy phenomenon of the fire plume. Height is the most important single dimension when the ceiling height is over 16 ft (4.9 m). The most sensitive detectors should be used when the ceiling height is over 30 ft (9 m). A ceiling height less than 10 ft (3 m) does not play a meaningful role in detector placement or detector response.

As ceiling height increases, spot-type heat detector spacing should be reduced. (See Table 9-1.)

Stratification: Stratification of air in a room can hinder air containing smoke particles or gaseous combustion products from reaching ceiling-mounted smoke or fire-gas detectors.

Stratification occurs when air containing smoke particles or gaseous combustion products is heated by smoldering or burning material and, becoming less dense than surrounding cooler air, rises until it reaches a level at which there is no longer a difference in temperature between it and the surrounding air.

In installations where detection of smoldering or small fires is desired and where the possibility of stratification exists, consideration should be given to mounting alternative detectors below the ceiling. (See Figure 9-1.) Specific designs for such an alternate detection system should be based upon an engineering survey.

TABLE 9-1. Detector Spacing and Ceiling Height

| Ceiling Height (ft) | | Percent of |
Above	Up To	Listed Spacing
0	10	100
10	12	91
12	14	84
14	16	77
16	18	71
18	20	64
20	22	58
22	24	52
24	26	46
26	28	40
28	30	34

NOTE: For a 30-ft (9-m) high ceiling, the listed spacing for spot-type heat detectors is reduced by two-thirds.

For SI Units: 1 ft = 0.30 m.

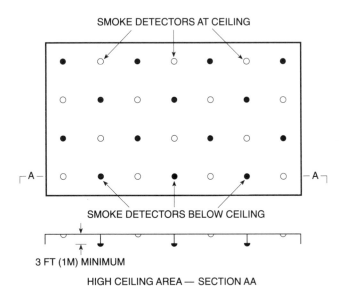

SMOKE DETECTORS AT CEILING

SMOKE DETECTORS BELOW CEILING

3 FT (1M) MINIMUM

HIGH CEILING AREA — SECTION AA

Figure 9-1. Detector location to overcome roof/ceiling air stratification. (For SI Units: 1 ft = 0.305 m.)

Ceiling Configuration: Ceiling configuration greatly influences the spacing of detectors and, therefore, the number of detectors needed. NFPA 72, *National Fire Alarm Code*, provides specific instructions on detector location related to the shapes of the ceilings. Level ceilings are those that are actually level or with a slope of 1.5 in. per ft (100 mm/m) or less. Sloped ceilings (of either the peaked or shed type) are defined as those with a slope greater than 1.5 in. per ft (100 mm/m). Ceiling structures can take many forms, such as the following.

• *Beam Construction:* These ceilings have solid structural or solid nonstructural members projecting down from the ceiling surface more than 4 in. (100 mm) and spaced more than 3 ft (0.9 m) on center.

• *Girders:* For these ceilings, girders support beams or joists and run at right angles to the beams or joists. When a girder is within 4 in. (100 mm) of the ceiling, it becomes a factor in determining the number of detectors and is to be considered as a beam. When the top of the girder is more than 4 in. (100 mm) from the ceiling, it is not a factor in detector location.

• *Solid Joist Construction:* These ceilings have solid structural or solid nonstructural members projecting down from the ceiling surface a distance of more than 4 in. (100 mm) and spaced at intervals 3 ft (0.9 m) or less, on center.

• *Smooth Ceiling:* This ceiling has a surface uninterrupted by continuous projections, such as solid joists, beams, or ducts extending more than 4 in. (100 mm) below the ceiling surface.

> NOTE: Open truss construction is not considered to impede the flow of fire products unless the upper member in continuous contact with the ceiling projects below the ceiling more than 4 in. (100 mm).

Room Ventilation: The effects of room ventilation on smoke detector placement is critical since smoke particles and fire gases are directly influenced by air movement. The higher the air velocity, the more complex the smoke detection design becomes. The heating, ventilating, and air conditioning (HVAC) systems incorporated in most facilities today are also affecting smoke detection system design. Smoke detectors should be positioned near return registers, not supply registers, since detectors mounted near supply registers are shielded from smoke by the incoming air supply. Detectors located near return registers, however, will be in the path of travel of smoke and fire gases. Detectors used in HVAC systems in accordance with NFPA 90A, *Standard for the Installation of Air Conditioning and Ventilating Systems*, should not be used instead of room detectors. Duct detectors, which shut down the HVAC system in the event of fire, should not be used as a substitute for open area protection because smoke may not enter the ductwork if the HVAC system is shut down. (See Figures 9-2 and 9-3.)

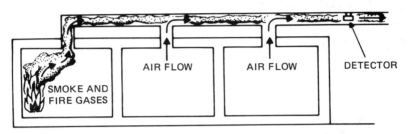

Figure 9-2. Dilution effect on operation of air duct detector.

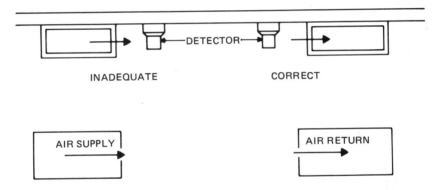

Figure 9-3. Smoke detector locations for ventilation systems.

Other design problems associated with HVAC systems and detector placement concern perforated membrane ceilings that evenly distribute supply air through a suspended ceiling. In the event of fire in these rooms, smoke and fire gases would not travel to the ceiling level until the fire plume developed hot thermal gases. One technique to avoid this problem is to use an air shield at least 3 sq ft (0.28 m^2) around the detector. (See Figures 9-4 and 9-5.)

For rooms and areas with high air movement, smoke detector spacing needs to be reduced. (See Figure 9-6 and Table 9-2.) The spacing coverage for high air movement should not be used for underfloor or

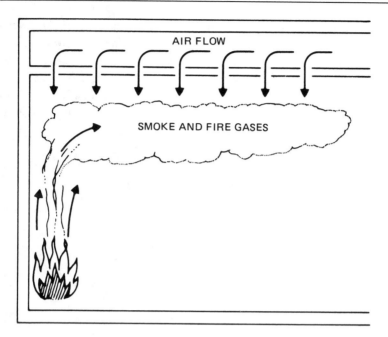

Figure 9-4. Air velocity barrier from ventilation system.

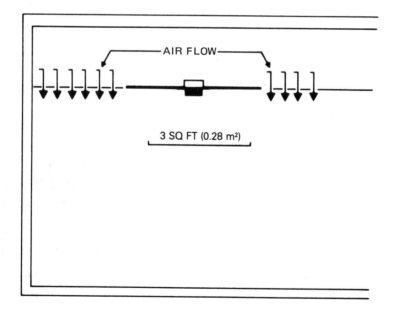

Figure 9-5. Air shield installation on perforated membrane ceiling system.

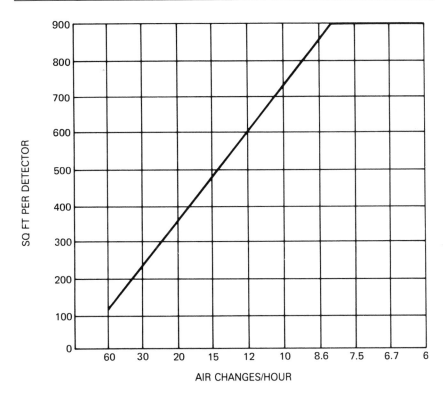

FOR S1 UNITS: 1 sq ft = 0.09 m².

Figure 9-6. Smoke detector spacing in high air movement areas.

TABLE 9-2. High Air Movement Areas

Minutes/ Air Change	Air Changes/ Hour	Sq Ft/ Detector
1	60.0	125
2	30.0	250
3	20.0	375
4	15.0	500
5	12.0	625
6	10.0	750
7	8.6	875
8	7.5	900
9	6.7	900
10	6.0	900

For SI Units: 1 sq ft = 0.0929 m².

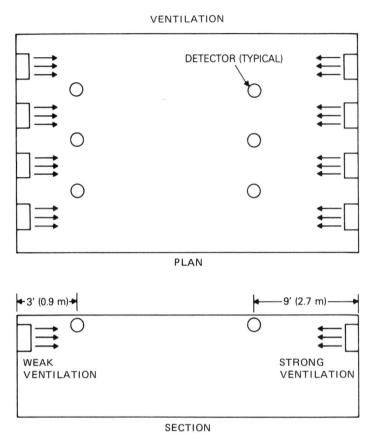

Figure 9-7. Detector placement in weak and strong ventilation areas.

above-ceiling spaces. Since the volume of these spaces is rather small (usually around ⅟₁₀th the volume of the room), reduced detector coverage is normally not required but may be influenced by physical construction and air movement within the space.

Detectors should be placed at least 3 ft (0.9 m) from an outlet opening in a weak ventilation area, and placed a minimum of 9 ft (2.7 m) from outlets in areas with strong ventilation. (See Figure 9-7.) Detectors must be placed so they are effective, regardless of the operating state of the ventilation system. (See Figure 9-8.) Rooms with exhaust ducts are best protected by a detector that is placed immediately after the last air exhaust opening (before the air enters a common duct), or dilution of the smoke could prevent early detector activation. (See Figure 9-9.)

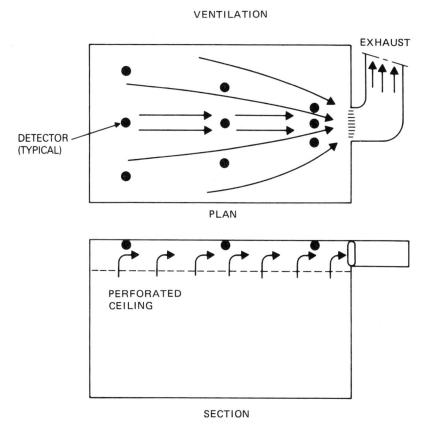

VENTILATION

EXHAUST

DETECTOR
(TYPICAL)

PLAN

PERFORATED
CEILING

SECTION

Figure 9-8. Detectors effective with HVAC system on or off.

Room Temperature: Room temperature is an important variable that must be determined in the design stage before detectors are selected. Many new detectors have solid-state electronic components that are adversely affected by extreme temperatures.

Smoke detectors are normally listed to be installed in areas where the ambient temperature is not more than 100°F (37.8°C) or less than 32°F (0°C). Control units fall within the same temperature range because of their sensitive electronic parts. During design, local heat sources, such as ovens, boilers, and furnaces, need to be considered relative to detector placement in the room.

Fixed temperature and rate compensated types of heat detectors are categorized by their operational temperature. (See Table 9-3.)

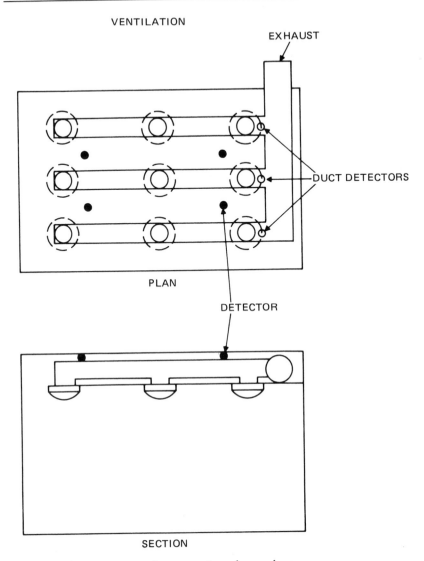

Figure 9-9. Detector placement in exhaust duct area.

Since most ceiling temperatures are under 100°F (37.8°C), ordinary temperature classification detectors are normally used. Local heat sources, including exhaust stacks, kilns, heaters, and lights, need to be considered in design of heat detector placement to guard against false alarms.

TABLE 9-3. Heat Detectors Classified by Operational Temperature

Temperature Classification	Temp. Rating Range (°F)	Max. Ceiling Temp. (°F)	Color Code
Low*	100 to 134	20 below†	Uncolored
Ordinary	135 to 174	100	Uncolored
Intermediate	175 to 249	150	White
High	250 to 324	225	Blue
Extra High	325 to 399	300	Red
Very Extra High	400 to 499	375	Green
Ultra High	500 to 575	475	Orange

*Intended only for installation in controlled ambient areas. Units marked to indicate maximum ambient installation temperature.

†Maximum ceiling temperature has to be 20°F or more below detector-rated temperature.

NOTE: The difference between the rated temperature and the maximum ambient temperature should be as small as possible to minimize the response time.

For SI Units: °C = 5/9 (°F -32).

HEAT DETECTORS

Heat detectors should be positioned on the ceiling not less than 4 in. (100 mm) from the sidewall, or on the sidewalls between 4 and 12 in. (100 and 300 mm) from the ceiling. Heat detectors should not normally be located on the lower flange of beams or suspended from the ceiling by conduit. Where beams are less than 12 in. (300 mm) in depth and less than 8 ft (2.4 m) on center, heat detectors can be installed on the bottoms of the beams. The space where the ceiling and side wall meet is normally considered a "dead air pocket" or "void." In room fire development, the concentration of fire heat and smoke could be low in this void space, so spot-type detectors should not be located here. (See Figure 9-10.)

Heat detectors placed on smooth ceilings should not exceed listed spacing unless permitted by an engineering analysis. The detectors should be within a distance of one-half the listed spacing, measured at a right angle, from all walls or partitions extending to within 18 in. (0.46 m) of the ceiling. For example: a rate compensation heat detector (UL listed) on a smooth ceiling should have 50-ft (15-m) spacing, and the spacing to a wall or partition should be 25 ft (7.6 m).

Heat detectors on a smooth ceiling should include all points of the ceiling within a distance equal to 0.7 times the listed spacing. (See Figure 9-11.)

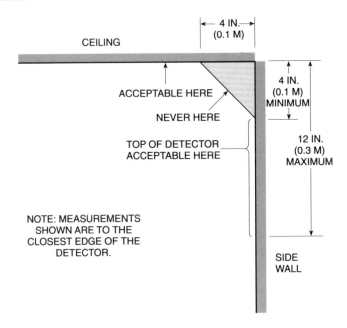

Figure 9-10. Spot-type detector locations.

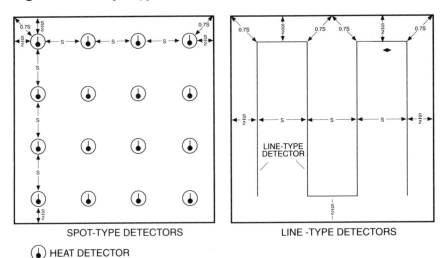

SPOT-TYPE DETECTORS   LINE -TYPE DETECTORS

(heat detector symbol) HEAT DETECTOR

S SPACING BETWEEN DETECTORS

Figure 9-11. Heat detector spacing for smooth ceiling.

Heat detectors placed on smooth ceilings in irregularly shaped areas can have greater than listed spacing between detectors, provided the maximum spacing from a detector to the farthest point on a side wall or corner is not greater than 0.7 times the listed spacing.

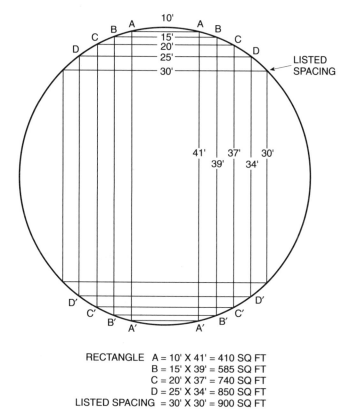

RECTANGLE A = 10' X 41' = 410 SQ FT
B = 15' X 39' = 585 SQ FT
C = 20' X 37' = 740 SQ FT
D = 25' X 34' = 850 SQ FT
LISTED SPACING = 30' X 30' = 900 SQ FT

Figure 9-12. Detector placement with listed 30-ft (9.1-m) spacing. (For SI Units: 1 ft = 0.30 m; 1 sq ft = 0.09 m².)

Figure 9-12 shows detector placement with listed 30-ft (9.1-m) spacing. A detector listed for a 30-ft (9.1-m) spacing will cover an area of a circle with a radius of 21 ft (6.4 m). For a rectangular area, a single properly located detector will meet the spacing requirement of the diagonal of the rectangle if the diagonal does not exceed the diameter of the circle. Detector efficiency will be increased due to the area of coverage being reduced.

In this case, 30 by 30 ft (9.1 × 9.1 m) is equal to 900 sq ft (83.6 m²), which is the maximum coverage for this detector. As an example, one detector would cover a corridor 10 by 41 ft (3 × 12.5 m) with the diagonal of the rectangle 21 ft (6.4 m) and a coverage of 410 sq ft (38 m²).

Figure 9-13 shows that spot-type heat detectors on solid joist construction should not exceed 50 percent of the smooth ceiling spacing when measured at right angles to the solid joists.

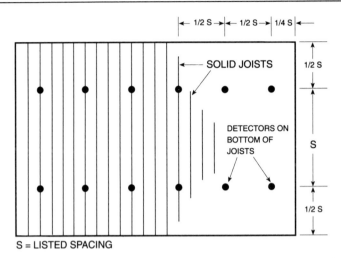

Figure 9-13. Spot-type heat detector spacing layouts for joisted ceiling. (S = listed spacing.)

Heat detector placement on beam construction is considered to be the same as for a smooth ceiling where the beams project no more than 4 in. (100 mm) below the ceiling. If the beams project more than 4 in. (100 mm) but less than 18 in. (0.46 m), the spacing of spot-type heat detectors at right angles to the direction of beam travel should not exceed two-thirds of the smooth ceiling spacing. If the beams project more than 18 in. (0.46 m) below the ceiling and are more than 8 ft (2.4 m) on center, each bay formed by the beams is considered to be a separate area.

Heat detectors placed on peaked ceilings should be located within 3 ft (0.9 m) of the peak of the ceiling, measured horizontally. (See Figure 9-14, bottom.) The spacing of additional detectors should be based on the horizontal projection of the ceiling in accordance with the type of ceiling construction. Placement is similar for detectors on shed-type ceilings, except that the detector should be positioned within 3 ft (0.9 m) of the highest side of the ceiling measured horizontally and spaced in accordance with the type of construction. (See Figure 9-14, top.)

SMOKE DETECTORS

Smoke detector spacing and location should be based on a detailed engineering study. Smoke development and travel are influenced by many parameters, including ceiling configuration, ceiling height, burning characteristics of materials, fuel arrangement, room geometry (including openings), and HVAC systems.

Since many variables affect detector placement, testing laboratories do not assign specific spacing to smoke detectors.

Spot-type smoke detectors should be placed on ceilings not less than 4 in. (100 mm) from any side wall to the near edge, or on the side wall between 4 and 12 in. (100 and 300 mm) down from the ceiling to the top of the detector. This positioning is similar to that for spot-type heat detectors.

Exceptions to positioning spot-type smoke detectors on the ceiling or side wall include use of detectors to mitigate smoke stratification, use of solid joist construction (with which a detector can be mounted on the bottom of the joist), and use of beam construction, for which a detector can be mounted on the lower beam flange if the beams are less than 12 in. (300 mm) deep and less than 8 ft (2.4 m) on center.

For projected beam-type smoke detectors, the beam should parallel the ceiling and be within 20 in. (0.5 m) of the ceiling. The listed length of the projected beam should be reduced by 33 percent for each mirror used. (See Table 9-4.)

In general, smoke detector spacing of 30 ft (9.1 m) is commonly used on smooth ceilings. Smoke detector spacing for solid joist construction with joists 8 in. (200 mm) or less is considered to be the same as that for

TABLE 9-4. Projected Beam Using Mirrors

Number of Mirrors	Maximum Allowable Beam Length	
O	Listed Length L	
1	2/3 L = a + b	
2	4/9 L = c + d + e	

EXAMPLE: Maximum allowable length of beam listed for 300 ft (L) using two mirrors is 4/9 × 300 ft, or 133 ft.

For SI Units: 1 ft = 0.30 m.

smooth ceilings. Spot-type smoke detector spacing perpendicular to the joist should be reduced if the joist is more than 8 in. (200 mm) deep.

Smoke detector spacing for beam construction with beams 8 in. (200 mm) or less is considered the same as that for smooth ceilings. Spacing of spot-type detectors should also be reduced perpendicular to the beams if beams exceed 8 in. (200 mm) in depth. Where beams are more than 18 in. (0.46 m) deep and more than 8 ft (2.4 m) on center, each bay should be considered a separate area containing at least one detector.

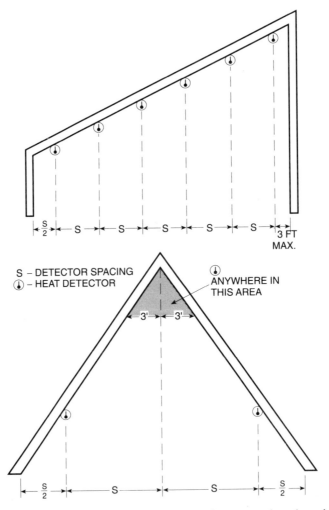

Figure 9-14. Heat and smoke detector placement for sloped ceilings (shed, top; peaked, bottom). (For SI Units: 1 ft = 0.30 m.)

Smoke detector placement for a sloped ceiling is the same as that for heat detectors. Smoke detectors for peaked ceilings should be located within 3 ft (0.9 m) of the peak, measured horizontally. Spacing for additional detectors is based upon the horizontal projection of the ceiling. (See Figure 9-14.)

Smoke detectors for shed-type ceilings should be mounted within 3 ft (0.9 m) of the highest side of the ceiling measured horizontally. Spacing of additional detectors is similar to that for peaked ceilings. (See Figure 9-14.)

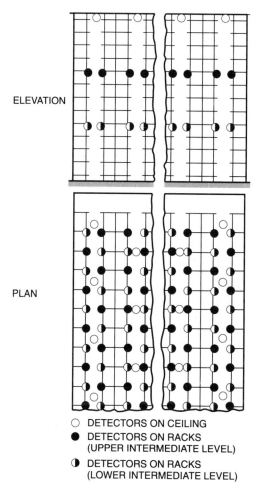

ELEVATION

PLAN

○ DETECTORS ON CEILING
● DETECTORS ON RACKS
  (UPPER INTERMEDIATE LEVEL)
◖ DETECTORS ON RACKS
  (LOWER INTERMEDIATE LEVEL)

Figure 9-15. Detector placement for typical closed rack storage.

Smoke detectors for high rack storage areas provide early warning of fire. For effective detection, smoke detectors should be positioned on the ceiling above each aisle and at intermediate levels in the racks. (See Figure 9-15.) Providing detectors at the ceiling level only may hamper early detection of fire, since the rack traps smoke which lacks the thermal lift to transport itself to the ceiling detectors.

FLAME-SENSING FIRE DETECTORS

Spacing for flame-sensing fire detectors should be based on engineering judgement in accordance with their listed spacing and the manufacturer's instructions. Since flame detectors respond to either infrared (IR) or ultraviolet (UV) radiation and essentially are line-of-sight devices, some of the following items might affect the operation of flame detectors: flickering light sources, sunlight, incandescent and fluorescent lights, and cutting and welding operations.

BIBLIOGRAPHY

NFPA Codes, Standards, Recommended Practices, and Manuals. (See the latest *NFPA Catalog* for availability of current editions of the following documents.)

NFPA 70, *National Electrical Code.*

NFPA 72, *National Fire Alarm Code.*

NFPA 90A, *Standard for the Installation of Air Conditioning and Ventilating Systems.*

Additional Reading

Heskestad, G., and Delichatsios, M.A., "Environments of Fire Detectors—Phase I: Effect of Fire Size, Ceiling Height, and Material," NBS-GCR77-95, The Fire Detection Institute Fire Test Report, Vol. II—"Analysis," National Institute for Standards and Technology, Gaithersburg, MD.

10

Heat and Smoke Detector Placement

INTRODUCTION

Appendix B to NFPA 72, *National Fire Alarm Code*, provides specific methodology for determining heat detector spacing based on the size and rate of growth of fire to be detected, various ceiling heights, and ambient temperature. The effects of ceiling height and the size and rate of growth of a flaming fire on smoke detector spacing are also addressed. It also proposes that a fire detection system can be designed to detect a given fire size of a specific heat release rate (Btu/sec). Appendix B utilizes the results of fire research funded by the Fire Detection Institute.

The data of Appendix B provide a method for modifying the listed spacing of both rate-of-rise and fixed temperature heat detectors required to achieve detector response to a geometrically growing flaming fire at a specific fire size. The data applicable to smoke detectors is limited to a theoretical analysis based on the flaming fire test data and is not intended to address the detection of smoldering fires.

For additional information on fire detector placement and design, see Chapter 9, "Fire Detection System Design"; Chapter 13, "Installation"; and Chapter 18, "Household Fire Warning Systems."

DETECTOR PLACEMENT METHODOLOGY

A method has been developed in NFPA 72, *National Fire Alarm Code*, to assist fire protection design engineers in the proper placement of fire detectors based on full-scale fire test results. These test results incorporate fire plume dynamics and the rate of heat release for a variety of combustible materials. This method can be used for designing the location of fire detectors, and to evaluate the adequacy of existing fire detection systems.

Design engineers should refer to NFPA 72, *National Fire Alarm Code*, Appendix B, for guidance in the proper analysis of detector selection and placement. Appendix B provides an analysis technique that is far superior to the out-dated method of symmetrically locating detectors on electrical drawings without regard to the building variables and fire development. Electrical drawings are typically not dimensioned, so the installing electrical contractor actually locates the detection devices, not the design engineer. With the new method, drawing dimensions are essential for optimum detector response, based upon the building features and detector response characteristics.

Fire research funded by the Fire Detection Institute provides the basic fire test data and analysis for the development of Appendix B. The fire tests conducted for the research were full-scale tests featuring geometrically growing flaming-mode fires. The research did not include smoldering-type fire scenarios, due to the complexity of predicting smoke particle movement without fire plume formation and ceiling jet flows. The understanding of smoke production, aging, agglomeration, and the transport mechanisms lags behind heat production. In addition, the smoke detector characteristics relating to specific fire environments are often not available to the design engineer. Therefore, smoke detector placement in Appendix B is for the detection of flaming-mode fires, rather than smoldering fires.

Limitations

Appendix B to NFPA 72, *National Fire Alarm Code*, has definite limitations for the placement of smoke detectors for early warning fire notification. If the intent of the design requires signal notification for incipient or smoldering-type fires, Appendix B should not be utilized. The design of early warning fire detection systems, with the goal of complete evacuation of the facility occupants prior to untenable conditions, is not addressed by Appendix B.

Appendix B is primarily used for property protection, such as the design of thermal detection systems that activate fire suppression systems. Facilities where Appendix B should *not* be applied include: computer rooms; control rooms; clean rooms; storage vaults; electrical rooms where smoldering fires are expected, such as switchgear rooms and motor control centers; and other area types where the anticipated design fire is incipient or smoldering.

MAIN VARIABLES

The primary purpose of Appendix B to NFPA 72, *National Fire Alarm Code*, is to achieve fire detector response at a specific threshold fire size based upon the rate of heat release. Key parameters affecting the detector response are the height of the ceiling (above the fire source), minimum ambient temperature, and the listed spacing of the detection devices. Knowledge of these parameters and an understanding of the methodology is all that is necessary to utilize Appendix B. The basic concept of Appendix B is that a fire detection system can be designed and installed to detect a given fire size of a specific heat release rate.

The five main variables addressed by Appendix B fire models are:

1. Time of fire growth,
2. Heat release rates,
3. Ceiling height,
4. Detector time constant (a function of the listed spacing), and
5. Listed spacing.

Time of fire growth and heat release rates will be discussed herein.

Time of Fire Growth

The time of fire growth is the time from flame initiation to the time fire develops a heat release rate of 1000 Btu/sec. Fires are categorized by their rate of growth as either being slow, medium, or fast developing. A slow-developing fire is one that takes at least 400 seconds or more to reach 1000 Btu/sec heat release rate. Not many fire scenarios fit the slow-developing fire classification.

A medium-developing fire is one that takes from 150 seconds to 400 seconds to reach 1000 Btu/sec rate of heat release. Examples of medium-developing fires are:

1. A ⅛-in. (3.2-mm) plywood wardrobe with fire-retardant interior finish has a fire growth time of 300 seconds.
2. A metal wardrobe with 94 lbs (42.6 kg) of combustible cartons has a fire growth time of 250 seconds.
3. A 36-lb (16.3 kg) metal chair frame with minimum cushion has a fire growth time of 200 seconds.

A fast-developing fire is one that takes less than 150 seconds to reach a 1000 Btu/sec rate of heat release. Examples of fast-developing fires are:

1. Polyethylene pallets stacked 3 ft (0.9 m) high have a fire growth time of 130 seconds.
2. A single, horizontal polyurethane foam mattress has a fire growth time of 110 seconds.
3. Wood pallets stacked 5 ft (1.5 m) high with a moisture content between 6 and 12 percent have a fire growth time between 90 and 190 seconds.

Figure 10-1 illustrates the time squared (t^2) type fire scenarios, which are characterized as slow, medium, or fast.

The 1000 Btu/sec heat release rate is an arbitrary figure. To put this number in perspective, the rate of heat release for listed thermal detectors that are tested with constant burning ethyl alcohol fires is approximately 1200 Btu/sec. The listed spacing of the detectors is compared to a 160°F (71°C) sprinkler on 10-ft (3-m) spacing.

Heat Release Rates

Over 100 calorimeter fire tests were performed on furniture and warehouse materials by Factory Mutual Research Corporation and the National Institute of Standards and Technology. These test results containing heat release rates, fire classification, fire intensity coefficient (α), and maximum heat release rates are listed in tables in Appendix B to NFPA 72, *National Fire Alarm Code*. Typical maximum rates of heat release from the Fire Detection Institute research are:

Medium wastebasket with milk cartons	100 Btu/sec
Large barrel with milk cartons	140 Btu/sec
Polyurethane foam upholstered chair	350 Btu/sec
Furnished living room (heat at door)	4000-8000 Btu/sec
Gasoline	200 Btu/sec/ft^2
Kerosene	200 Btu/sec/ft^2
Diesel oil	180 Btu/sec/ft^2
Methyl alcohol	65 Btu/sec/ft^2

Note that the rates of heat release for pool fires of flammable and combustible liquids have a steady-state heat release rate. These pool fires do not follow the t^2-type fire scenario. If Appendix B is used as a guide for pool fires, time to actuation of the heat detector will be less in the pool fire than the calculated fire growth time.

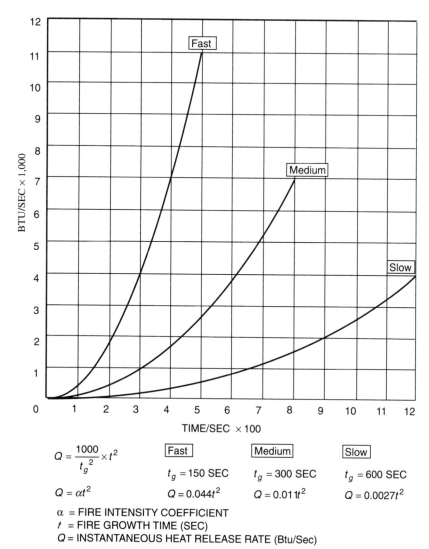

Figure 10-1. Graphic comparison of slow-, medium-, and fast-developing fires.

Threshold Fire Size: From a design perspective, a threshold fire size, Q_d, is selected by the engineer as the rate of heat release at which detection is desired. This selection needs to be determined on the level of desired protection. The detection system is designed and installed to detect a threshold fire size at a specific heat release rate.

For example, a design engineer is protecting a storage tank of kerosene and wants to detect a 5 sq ft (0.45 m²) pool fire. Since kerosene liberates a maximum heat release rate of 200 Btu/sec/ft^2, the rate of heat release of the liquid for the 5 sq ft (0.45 m^2) area is 1000 Btu/sec. This would be Q_d, the threshold fire size for the anticipated pool fire.

Fire Growth Model: The design fires used in Appendix B to NFPA 72, *National Fire Alarm Code*, grow according to the following equation:

$$Q = \left(1000/t_s^2\right)t^2$$

where

Q   = Heat release rate (Btu/sec)
t_g = Fire growth time (slow, medium, or fast) (sec)
t   = Time after flaming occurs (sec)
$1000/t_g^2$ = Fire intensity coefficient (α)

This t^2 fire formula is the mathematical model that forms the foundation of Appendix B.

The equation for the rise in temperature of a heat detector is expressed by:

$$\frac{\Delta T_s}{dt} = \frac{\left(T_g - T_d\right)}{\tau}$$

where

T_d = Temperature rating of the detector
T_g = Gas temperature at the detector
τ = Detector time constant.

The detector time constant (τ) is a measure of the sensitivity characteristics of the particular detector. The detector time constant is measured in the Factory Mutual Research Corporation plunge test apparatus at a reference velocity of 5 ft/sec (1.5 m/sec). The equation is expressed as:

$$\tau = \left(mc\right)/\left(hA\right)$$

where
(τ) = Detector time constant
M = Detector element mass
C = Detector element specific heat
h = Convective heat transfer coefficient
A = Surface area of the detector element.

The convective heat transfer coefficient, h, varies as the square root of the gas velocity, U, as the gas passes the detector.

Response Time Index: The response time index (RTI) is equal to the detector time constant, τ, multiplied by the square root of the gas velocity referenced at 5 ft/sec (1.5 m/sec).

$$RTI = \tau \sqrt{u}$$

where
RTI = Response time index ($sec^{1/2}$ $ft^{1/2}$)
τ = Detector time constant (sec)
U   = Gas velocity set at 5 ft/sec (1.5 m/sec).

A detector having a RTI of 56 $sec^{1/2}$ $ft^{1/2}$ has a detector time constant (τ) of 25 sec, when measured at a gas velocity of 5 ft/sec (1.5 m/sec).

Spacing of Underwriters Laboratories Inc. listed or Factory Mutual Research Corporation approved heat detectors is based on the detector time constant, τ, as established in Appendix B to NFPA 72, *National Fire Alarm Code*. (See Table 10-1.)

TABLE 10-1. Time Constants (τ_0) for Any Listed Heat Detector*

Listed Spacing (ft)	UL						FMRC All Temps.
	128°F	135°F	145°F	160°F	170°F	196°F	
10	400	330	262	195	160	97	196
15	250	190	156	110	89	45	110
20	165	135	105	70	52	17	70
25	124	100	78	48	32		48
30	95	80	61	36	22		36
40	71	57	41	18			
50	59	44	30				
70	36	24	9				

NOTE 1: These time constants are based on an analysis of the Underwriters Laboratories Inc. (UL) and Factory Mutual Research Corporation (FMRC) listing test procedures. Plunge test results performed on the detector to be used will give a more accurate time constant. See Appendix B to NFPA 72, *National Fire Alarm Code*, for a further discussion of detector time constants.

NOTE 2: These time constants can be converted to response time index (RTI) values by multiplying by $\sqrt{5}$ ft/sec ($\sqrt{1.5}$ m/s). (See Appendix B to NFPA 72, *National Fire Alarm Code*.)

*At a reference velocity of 5 ft/sec (1.5 m/s).

For SI Units: 1 ft = 0.30 m; °C = 5/9 (°F–32).

As an example, if the temperature rating of a heat detector is 145°F (63°C), and the UL-listed spacing is 50 ft (15 m), then the detector time constant, τ, is 30 sec. Then,

$$\mathrm{RTI} = \tau(U)^{1/2} = 30(5)^{1/2} = 67 \ \mathrm{sec}^{1/2} \ \mathrm{ft}^{1/2}$$

Selection Method for Flaming-Mode Fire Detector Placement

1. Establish the goals of the fire detection system.
2. Select the threshold fire size, Q_d, which is the desired rate of heat release at the time of detection.
3. Select the fire growth time, t_g, which is the detection time and response time.
4. Determine the environmental conditions, including ceiling height and minimum ambient ceiling temperature.
5. Select the fire detector to be used, including type and sensitivity.
6. Use the tables from Appendix B to NFPA 72, *National Fire Alarm Code*, to determine the design spacing.
7. Evaluate the design choice.
8. Re-design and calculate, if necessary.

Example Fire Scenarios

EXAMPLE FOR FIXED THERMAL DETECTOR:

A warehouse containing combustible materials [wood pallets stacked 1.5 ft (0.45 m) high and having an area of approximately 10 sq ft (0.9 m²)] will be provided with a pre-action sprinkler system. The ceiling height is 16 ft (4.8 m) above the floor, and the minimum ambient temperature at the ceiling level is 55°F (13°C). Determine the heat detector spacing with a threshold fire size of 1000 Btu/sec, if the desired time of detection (fire growth time) is 150 seconds (2.5 minutes after flaming ignition occurs), and a fixed temperature 145°F (63°C) detector UL listed for 50-ft (15-m) spacing is used.

Input Data:

Threshold fire size:	1000 Btu/sec
Fire growth time:	150 sec
Ceiling height:	16 ft (4.8 m)
Minimum ambient temperature:	55°F (13°C)
Detector:	Fixed temperature 145°F (63°C) UL listed for 50 ft (15 m).

SOLUTION:

The fire intensity coefficient (α) =

$$\frac{1000}{t_g^2} = \frac{1000}{(150)^2} = 0.044 \text{ Btu/Sec}^2$$

From Table 10-1, the detector time constant (τ) for any listed detector (sec): For a 50-ft (15-m) listed spacing and a fixed temperature of 145°F (63°C), the time constant is 30.

The response time index (RTI) is 30 times the square root of 5 ft/sec (1.5 m/sec) velocity from the Factory Mutual Research Corporation plunge test.

$$\text{RTI} = 30 \times (5)^{1/2} = 67 \text{ ft}^{1/2} \text{ sec}^{1/2}$$

The change in temperature during the fire is:

$$\Delta T = T_s - T_o = 145 - 65 = 80°F$$

From Table 10-2 the detector spacing is between:
For τ = 25 sec: spacing = 16 ft (4.8 m).
For τ = 50 sec: spacing = 12 ft (3.6 m).
By interpolation: 16 − [25-30 (16−12/25−50)] = 15.2 ft (4.6 m).

The spacing for the fixed temperature detectors should be designed for 15.2 ft (4.6 m).

EXAMPLE FOR FIXED THERMAL DETECTOR:

A process facility utilizing polystyrene is planned to be protected with a deluge sprinkler system. A pilot head pneumatic detection system will activate the deluge valve. UL-listed spacing for the heat detector is 30 ft (9 m), with a temperature rating of 170°F (77°C). The ceiling height is 12 ft (3.6 m), and the minimum ambient temperature is 70°F (21°C). Determine the detector spacing for a fire growth time of 50 sec (desired time of detection) for a threshold fire size of 1000 Btu/sec.

Input Data:

Threshold fire size:	1000 Btu/sec
Fire growth time:	50 sec
Ceiling height:	12 ft (3.6 m)
Minimum ambient temperature:	70°F (21°C)
Detector:	Fixed temperature 170°F (77°C) UL listed for 30 ft (9 m).

TABLE 10-2. Threshold Fire Size at Response: 1000 Btu/sec
Fire Growth Rate: 150 sec to 1000 Btu/sec
α: 0.044 Btu/sec³

τ	RTI	ΔT	CEILING HEIGHT IN FEET						
			4.0	8.0	12.0	16.0	20.0	24.0	28.0
			INSTALLED SPACING OF DETECTORS						
25	56	40	39	36	32	28	25	21	17
25	56	60	32	28	24	21	17	13	9
25	56	80	27	24	20	16	12	8	4
25	56	100	24	20	16	12	8	4	1
25	56	120	22	18	14	10	6	2	0
25	56	140	20	16	12	8	4	0	0
50	112	40	31	29	26	22	19	16	13
50	112	60	25	23	19	16	13	9	6
50	112	80	22	19	15	12	9	6	3
50	112	100	19	16	13	9	6	3	0
50	112	120	17	14	11	7	4	1	0
50	112	140	16	13	9	6	2	0	0

NOTE: Detector time constant at a reference velocity of 5 ft/sec.
For SI Units: 1 ft = 0.30 m; 1000 Btu/sec = 1055 kW.

SOLUTION:

The fire intensity coefficient (α) =

$$\frac{1000}{(t_g)^2} = \frac{1000}{(50)^2} = 0.40 \text{ Btu/sec}^3$$

From Table 10-1, the detector time constant (τ) for any listed detector (sec): For a 30-ft (9-m) listed spacing and a fixed temperature of 170°F (77°C), the time constant is 22 sec. (Use τ of 25 sec.)

The response time index (RTI) is 22 times the square root of 5 ft/sec (1.5 m/sec) velocity from the Factory Mutual Research Corporation plunge test.

$$RTI = 22 \times (5)^{1/2} = 49 \text{ ft}^{1/2} \text{ sec}^{1/2}$$

The change in temperature during the fire is:

$$\Delta T = T_s - T_o = 170 - 70 = 100°F$$

From Table 10-3, the detector spacing is between:
For τ = 25 sec: Δ T = 100°F; spacing = 8 ft (2.4 m).
Therefore, for a τ = 25 sec, and the Δ T = 100°F (37.8°C), then the design spacing is 12 ft (3.6 m).

TABLE 10-3. Threshold Fire Size at Response: 750 Btu/sec
Fire Growth Rate: 150 sec to 1000 Btu/sec
α: 0.044 Btu/sec^3

τ	RTI	ΔT	CEILING HEIGHT IN FEET						
			4.0	8.0	12.0	16.0	20.0	24.0	28.0
			INSTALLED SPACING OF DETECTORS						
25	56	40	32	29	26	22	18	15	11
25	56	60	26	23	19	15	12	8	4
25	56	80	23	19	15	11	8	4	1
25	56	100	20	16	12	8	5	1	0
25	56	120	18	14	10	6	3	0	0
25	56	140	16	12	8	5	1	0	0
50	112	40	25	23	20	17	14	11	8
50	112	60	21	18	15	12	8	5	3
50	112	80	18	15	12	8	5	2	0
50	112	100	16	13	9	6	3	1	0
50	112	120	14	11	8	4	2	0	0
50	112	140	13	10	6	3	1	0	0
75	168	40	22	19	17	14	11	8	5
75	168	60	18	15	12	9	6	4	1
75	168	80	15	12	9	6	4	1	0
75	168	100	13	10	7	5	2	0	0
75	168	120	12	9	6	3	1	0	0
75	168	140	11	8	5	2	0	0	0
100	224	40	19	17	14	12	9	6	4
100	224	60	16	13	10	8	5	3	1
100	224	80	13	11	8	5	3	1	0
100	224	100	12	9	6	3	1	0	0
100	224	120	10	8	5	2	1	0	0
100	224	140	9	7	4	1	0	0	0
125	280	40	17	15	13	10	7	5	3
125	280	60	14	12	9	6	4	2	1
125	280	80	12	9	7	4	2	0	0
125	280	100	10	8	5	3	1	0	0
125	280	120	9	7	4	2	0	0	0
125	280	140	8	6	3	1	0	0	0
150	335	40	16	14	11	9	6	4	2
150	335	60	13	10	8	5	3	1	0
150	335	80	11	8	6	3	1	0	0
150	335	100	9	7	4	2	1	0	0
150	335	120	8	6	3	1	0	0	0
150	335	140	8	5	3	1	0	0	0

NOTE: Detector time constant at a reference velocity of 5 ft/sec.

For SI Units: 1 ft = 0.30 m; 1000 Btu/sec = 1055 kW.

TABLE 10-3. Threshold Fire Size at Response: 750 Btu/sec
Fire Growth Rate: 150 sec to 1000 Btu/sec
α: 0.044 Btu/sec³ (continued)

τ	RTI	ΔT	CEILING HEIGHT IN FEET						
			4.0	8.0	12.0	16.0	20.0	24.0	28.0
			INSTALLED SPACING OF DETECTORS						
175	391	40	15	13	10	8	5	3	2
175	391	60	12	9	7	5	2	1	0
175	391	80	10	8	5	3	1	0	0
175	391	100	9	6	4	2	0	0	0
175	391	120	8	5	3	1	0	0	0
175	391	140	7	4	2	0	0	0	0
200	447	40	14	12	9	7	4	3	1
200	447	60	11	9	6	4	2	1	0
200	447	80	9	7	4	2	1	0	0
200	447	100	8	6	3	1	0	0	0
200	447	120	7	5	2	1	0	0	0
200	447	140	6	4	2	0	0	0	0
225	503	40	13	11	8	6	4	2	1
225	503	60	10	8	6	3	2	0	0
225	503	80	9	6	4	2	0	0	0
225	503	100	8	5	3	1	0	0	0
225	503	120	7	4	2	0	0	0	0
225	503	140	6	4	1	0	0	0	0
250	559	40	12	10	8	5	3	2	0
250	559	60	10	7	5	3	1	0	0
250	559	80	8	6	4	2	0	0	0
250	559	100	7	5	2	1	0	0	0
250	559	120	6	4	2	0	0	0	0
250	559	140	6	3	1	0	0	0	0
275	615	40	12	10	7	5	3	1	0
275	615	60	9	7	5	3	1	0	0
275	615	80	8	5	3	1	0	0	0
275	615	100	7	4	2	1	0	0	0
275	615	120	6	4	2	0	0	0	0
275	615	140	5	3	1	0	0	0	0
300	671	40	11	9	7	4	3	1	0
300	671	60	9	7	4	2	1	0	0
300	671	80	7	5	3	1	0	0	0
300	671	100	6	4	2	0	0	0	0
300	671	120	6	3	1	0	0	0	0
300	671	140	5	3	1	0	0	0	0

NOTE: Detector time constant at a reference velocity of 5 ft/sec.
For SI Units: 1 ft = 0.30 m; 1000 Btu/sec = 1055 kW.

TABLE 10-3. Threshold Fire Size at Response: 750 Btu/sec
Fire Growth Rate: 150 sec to 1000 Btu/sec
α: 0.044 Btu/sec^3 (continued)

			CEILING HEIGHT IN FEET						
τ	RTI	ΔT	4.0	8.0	12.0	16.0	20.0	24.0	28.0
			INSTALLED SPACING OF DETECTORS						
325	727	40	11	9	6	4	2	1	0
325	727	60	9	6	4	2	1	0	0
325	727	80	7	5	3	1	0	0	0
325	727	100	6	4	2	0	0	0	0
325	727	120	5	3	1	0	0	0	0
325	727	140	5	2	1	0	0	0	0
350	783	40	10	8	6	4	2	1	0
350	783	60	8	6	4	2	0	0	0
350	783	80	7	4	2	1	0	0	0
350	783	100	6	3	2	0	0	0	0
350	783	120	5	3	1	0	0	0	0
350	783	140	5	2	0	0	0	0	0
375	839	40	10	8	5	3	2	0	0
375	839	60	8	6	3	2	0	0	0
375	839	80	6	4	2	0	0	0	0
375	839	100	6	3	1	0	0	0	0
375	839	120	5	3	1	0	0	0	0
375	839	140	4	2	0	0	0	0	0
400	894	40	10	7	5	3	2	0	0
400	894	60	8	5	3	1	0	0	0
400	894	80	6	4	2	0	0	0	0
400	894	100	5	3	1	0	0	0	0
400	894	120	5	2	1	0	0	0	0
400	894	140	4	2	0	0	0	0	0

NOTE: Detector time constant at a reference velocity of 5 ft/sec.
For SI Units: 1 ft = 0.30 m; 1000 Btu/sec = 1055 kW.

EXAMPLE FOR RATE-OF-RISE THERMAL DETECTOR:

For the same problem as stated in the example for a fixed thermal detector, but with rate-of-rise heat detectors instead of fixed temperature devices, the solution is as follows.

Input Data:

Threshold fire size:	1000 Btu/sec
Fire growth time:	Fast
Ceiling height:	12 ft (3.6 m)
Detector:	Rate-of-rise UL listed for 30 ft (9 m).

SOLUTION:

From Table 10-4, for a 30-ft (9-m) listed spacing rate-of-rise detector at a ceiling height of 12 ft (3.6 m), using a threshold fire size of 1000 Btu/sec, the design spacing is 28 ft (8.4 m).

Since Table 10-4 is based upon UL-listed spacing of 50 ft (15 m), a modifier must be used for the UL-listed spacing of 30 ft (9 m). From Table 10-5, the modifier is 0.85 for a "fast" fire growth rate. Therefore,

Design spacing = 0.85 × 28 ft = 23.8 ft (7.14 m).

Other design elements being equal, the design engineer should use the rate-of-rise detector having a design spacing of 23.8 ft (7.14 m), instead of the fixed temperature device that has a design spacing of 8 ft (2.4 m).

EXAMPLE FOR RATE-OF-RISE DETECTOR:

An electrical vault housing electrical transformers containing combustible oil is planned to be provided with a fixed water spray system. The water spray system will be activated by rate-of-rise thermal detectors having a UL-listed spacing of 50 ft (15 m). The ceiling height of the vault is 15 ft (4.5 m). The fire growth time is "fast," and the threshold fire size for detection is 500 Btu/sec.

Input Data:

Threshold fire size:	500 Btu/sec
Fire growth time:	Fast
Ceiling height:	15 ft (4.5 m)
Detector:	Rate-of-rise UL listed for 50 ft (15 m).

If the rate of heat release for the combustible oil is approximately 200 Btu/sec/ft^2, then the threshold fire size of 500 Btu/sec is approximately 2.5 ft^2 (0.23 m^2).

SOLUTION:

From Table 10-4, for a 50-ft (15-m) listed spacing rate-of-rise detector at a ceiling height of 15 ft (4.5 m), using a threshold fire size of 500 Btu/sec, the design spacing is 17 ft (5.1 m). Therefore, the spacing for the rate-of-rise detectors should be designed for 17 ft (5.1 m).

If the UL listing is not 50 ft (15 m), Table 10-5 must be used to modify the spacing. As an example, if the UL-listed spacing is 30 ft (9 m), the design spacing of 17 ft (5.1 m) would be required to be multiplied by 0.85. The designed spacing would change to 14.5 ft (4.35 m).

TABLE 10-4. Installed Spacings for Rate-of-Rise Heat Detectors (Threshold Fire Size and Growth Rate)

Ceiling Height (ft)	$Q_d = 1000$ Btu/sec			$Q_d = 750$ Btu/sec			$Q_d = 500$ Btu/sec			$Q_d = 250$ Btu/sec			$Q_d = 100$ Btu/sec		
	s	m	f	s	m	f	s	m	f	s	m	f	s	m	f
4	28	32	32	26	28	27	22	24	23	16	17	16	11	11	10
5	27	31	31	25	27	27	21	23	22	15	16	15	10	10	9
6	26	30	31	24	26	27	20	22	22	15	15	15	9	9	9
7	25	29	30	23	26	26	19	21	21	14	14	14	9	9	8
8	24	29	30	22	25	26	18	21	21	13	13	14	8	8	8
9	23	28	29	21	24	25	17	20	20	12	13	13	7	7	7
10	22	27	29	20	23	25	16	19	20	12	12	13	7	7	7
11	21	27	28	18	23	24	15	19	19	11	12	12	6	6	6
12	20	26	28	17	22	24	15	18	19	10	11	12	5	5	5
13	19	25	27	16	22	23	14	18	18	9	11	11	5	5	5
14	18	24	27	15	21	22	13	17	18	9	10	11	4		
15	16	24	26	14	20	21	12	17	17	8	10	10			
16	15	23	25	13	19	21	11	16	16	7	9	10			
17	14	22	25	12	19	20	10	15	16	6	9	9			
18	13	22	24	11	18	20	9	14	15		8	8			
19	12	21	23	10	17	19	8	14	14		8	8			
20	11	20		9	16	19	7	13	14		7	7			
21	10	19		8	15	18		12	13		7				
22	9	19		7	15	17		12	13		6				
23	8	18			14	17		11	12		5				
24		17			13	16		11	11		5				
25		16			12	15		10	10		4				
26		15			12	15		9	10						
27		14			11	14		9							
28		13			11	13		8							
29		13			10			8							
30		12			10			7							

s = "slow" fire, m = "medium" fire, f = "fast" fire.

TABLE 10-5. Spacing Modifiers for Rate-of-Rise Heat Detectors

Listed Spacing (ft)	(Fire Growth Rate)		
	Slow	Medium	Fast
15	0.57	0.55	0.45
20	0.72	0.63	0.62
25	0.84	0.78	0.76
30	0.92	0.86	0.85
40	0.98	0.96	0.95
50	1.00	1.00	1.00
70	1.00	1.01	1.02

For SI Units: 1 ft = 0.30 m.

EXAMPLE FOR SMOKE DETECTORS FOR FLAMING-MODE-TYPE FIRES:

A storage area containing combustible materials will be provided with a pre-action sprinkler system. The ceiling height is 22 ft (6.6 m) above the floor. Determine the smoke detector spacing with a threshold fire size of 750 Btu/sec, with a "medium" fire growth rate.

Input Data:

Threshold fire size:	750 Btu/sec
Fire growth time:	Medium
Ceiling height:	22 ft (6.6 m).

SOLUTION:

From Figure 10-2, for a threshold fire size of 750 Btu/sec and with a ceiling height of 22 ft (6.6 m), the installed detector design spacing is 40 ft (12 m).

EXAMPLE FOR EVALUATING EXISTING THERMAL DETECTOR PLACEMENT:

A fabricating area has 135°F (57°C) fixed temperature detectors listed for 50-ft (15-m) spacing. The actual spacing on the 16-ft (4.8-m) high ceiling is 30 ft (9 m), and the minimum ambient temperature at the ceiling level is 55°F (12.8°C). The anticipated fire growth time is 150 sec. Determine the threshold fire size, Q_d.

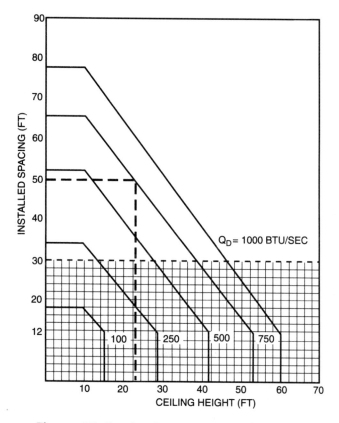

Figure 10-2. Smoke Detector, Medium Fire.

Input Data:

Fire growth time:	150 sec
Ceiling height:	16 ft (4.8 m)
Minimum ambient temperature:	55°F (12.8°C)
Detector:	Fixed temperature 135°F (57°C) UL listed for 50 ft (15 m).

SOLUTION:

The fire intensity coefficient (α) =

$$\frac{1000}{\left(t_g\right)^2} = \frac{1000}{\left(150\right)^2} = -.044 \text{ Btu/sec}^3$$

From Table 10-1 the detector time constant, τ, is 44 sec.

The response time index (RTI) is 44 times the square root of 5 ft/sec (1.5 m/sec) velocity from the Factory Mutual Research Corporation plunge test.

$$\text{RTI} = 44 \times (5)^{1/2} = 98 \text{ ft}^{1/2} \text{ sec}^{1/2}$$

The change in temperature during a fire is:

$$\Delta T = 135 - 55 = 80°F$$

From Table 10-6, the threshold fire size, Q_d, is between:

For $\tau = 25$ sec: $\Delta T = 80°F$; fire size = 2088 Btu/sec.

For $\tau = 50$ sec: $\Delta T = 80°F$; fire size = 2563 Btu/sec.

By interpolation:

$Q_d = 2088 - [(25\text{-}44)\ (2088\text{-}2563)/(25\text{-}50)] = 2449$ Btu/sec.

Therefore, for a $\tau = 44$ sec, and the $\Delta T = 80°F$ (26.7°C), the threshold fire size, Q_d, is 2449 Btu/sec.

TABLE 10-6. Installed Spacing of Heat Detector: 30 ft
Fire Growth Rate: 150 sec to 1000 Btu/sec
α: 0.044 Btu/sec³

			CEILING HEIGHT IN FEET						
τ	RTI	ΔT	4.0	8.0	12.0	16.0	20.0	24.0	28.0
			FIRE SIZE AT DETECTOR RESPONSE (Btu / sec)						
25	56	40	684	780	911	1066	1243	1443	1665
25	56	60	919	1084	1300	1556	1850	2183	2555
25	56	80	1150	1399	1715	2088	2518	3005	3550
25	56	100	1387	1728	2157	2663	3245	3904	4641
25	56	120	1629	2072	2626	3278	4026	4872	5819
25	56	140	1877	2432	3122	3931	4858	5906	7077
50	112	40	958	1073	1220	1391	1583	1796	2030
50	112	60	1284	1465	1696	1965	2270	2613	2992
50	112	80	1603	1856	2182	2563	3000	3492	4042
50	112	100	1919	2251	2683	3192	3775	4436	5177
50	112	120	2217	2654	3203	3852	4598	5445	6394
50	112	140	2526	3066	3743	4544	5467	6514	7688

NOTE: Detector time constant at a reference velocity of 5 ft/sec.
For SI Units: 1 ft = 0.30 m; 1000 Btu/sec = 1055 kW.

BIBLIOGRAPHY

NFPA Codes, Standards, Recommended Practices, and Manuals. (See the latest *NFPA Catalog* for availability of current editions of the following documents.)

NFPA 72, *National Fire Alarm Code*.

Additional Reading

DiNenno, P. J., ed., *SFPE Handbook of Fire Protection Engineering*, 1st ed., National Fire Protection Association, Quincy, MA, 1988.

11

Engineering Documents

INTRODUCTION

Preparation of engineering documents is a major aspect in the design and installation of a fire alarm signaling system. Engineering documents, which are generally divided into drawings and specifications, must be prepared accurately, precisely, and completely for all components of a fire alarm signaling system to operate properly. Clear drawings and carefully worded specifications will ensure that the facility owner, contractor, and the fire alarm signaling system designer work toward this end. Additional information on design criteria that affect engineering documents can be found in Chapter 9, "Fire Detection System Design."

Conceptual engineering documents concerning system design should include defined objectives, goals, and criteria. The design objectives outline basic functions, such as life safety, actuating suppression systems, operating emergency building features, etc., that the system must meet to be effective.

The design criteria describes the specific parameters of the fire detection system and become a basic document for design drawings and specifications. The design criteria evaluate each area of the facility being protected. Building features analyzed on an area-by-area basis should include type of combustibles (to postulated fire scenarios), ceiling height and configuration, stratification, room ventilation and temperature, implications for automatic fire detectors, and detector selection and spacing.

After design criteria have been developed, the principal documents prepared for the installation of fire alarm signaling systems should include drawings and specifications that contain standard clauses and technical instructions.

These documents are the construction contract, which governs the rights, duties, and liabilities of the contractor or vendor who performs the work, the person or organization for whom the work is to be executed, and the engineers who design, inspect and test the work.

DESIGN DRAWINGS

Design drawings provide the contractor with needed information in easily understood pictorial form. Drawings must be clear and accurate, since carelessly prepared or inadequate drawings can cause trouble in the performance of a contract and are a prime source of a contractor's claims for extra compensation. Engineers should provide sufficient time and personnel to prepare suitable drawings. If initial contract drawings for a fire detection system are incomplete, any features added to the system by the owner or engineer will appreciably increase system cost. Some claims by the contractor to recover these extra costs are inevitable and usually justifiable. This should be understood by the owner and engineer if the proposal drawings are not thorough and comprehensive.

In general, engineering and architectural drawings (or "plans") are called design drawings when they are prepared for bidding purposes. Once the contract is awarded, design drawings become known as contract drawings.

Basically, design drawings show the general outline and sufficient detail regarding the system to enable the contractor to "bid" the job. These drawings seldom are entirely adequate for every need connected with the installation of the fire detection system. Engineering drawings are often supplemented with detail drawings, shop drawings, construction drawings, or working drawings that are prepared by or for the contractor. A detailed drawing, for example, could be a schematic of the electrical circuitry inside a fire alarm and detection system control unit.

Types of Drawings

Four types of drawings can generally depict any size fire alarm signaling system. These are:

1. Equipment Location Drawings. These drawings are used on large fire alarm signaling systems to graphically show the location of the main annunciator, remote annunciators, printers, control units, transmitters, transponders, and municipal fire alarm system connections at the facility.

2. Elementary Wiring Diagram (Riser Diagram). These diagrams illustrate the arrangement of fire detection and alarm boxes related to control units. Typical data included on these diagrams consist of control panel or unit, power supply input, alarm initiating and alarm indicating circuits, auxiliary output function (such as suppression system actuation, HVAC system shutdown, elevator recall systems), etc. (See Figure 11-1.)

3. Fire Detector Location Drawings. The key element of these drawings is the location dimensioning of each fire detector and alarm box. These drawings should be prepared by persons familiar with NFPA 72, *National Fire Alarm Code*. The following data should be available before these drawings are completed:

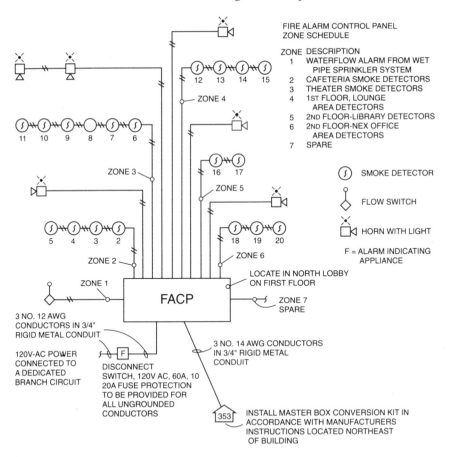

Figure 11-1. Typical fire alarm riser diagram.

(a) Ceiling construction (structural drawings)
- smooth ceiling
- beam construction
- joist construction
- slope construction

(b) HVAC system
- air supply and return location for each area having fire detection
- HVAC system controls (see NFPA 90A, *Standard for the Installation of Air Conditioning and Ventilating Systems*)

(c) Architectural drawings
- room dimensions
- sections showing ceiling height
- equipment locations.

If an electrical contractor prepares the drawings that detail fire detector location, the contractor should consult the appropriate code or standard in force. Without proper detector dimensions on the drawings, a detector can be easily installed in an inappropriate location (e.g., on a lower flange or truss of a beam or adjacent to HVAC equipment).

4. Conduit and Wiring Drawings. These drawings identify the conduit and wiring for field-routed junction boxes, control units, and detection system devices. Conduit is normally not dimensioned on drawings but is located while the building is under construction.

Preparing Drawings

The following points should be included in the contract if the contractor is to prepare drawings:

1. The size of drawings. [If the drawings are to be microfilmed, a minimum scale of $\frac{1}{4}$ in. = 1 ft (6.35 mm = 0.30 m) should be required.]

2. Instructions. Use of paper, mylar, or acetate for tracings, pencil or ink for the final drawings, and printed forms should be determined.

3. The general title and title block to be used for all drawings pertaining to the project.

4. The printed forms to be used on the drawings for recording revisions and approval by the engineer.

5. Statement concerning responsibility for the accuracy of drawings.

6. The need and procedure for the submission of any preliminary layouts for approval by the engineer.
7. The number of prints required and the procedure for submitting shop drawings for the approval of the engineer.
8. The number of prints required for filing purposes by the owner and the engineer, and the number of prints to be sent to the inspectors and to the engineer's field staff.

SPECIFICATIONS

Specifications are important since they are legal contractual documents. Specifications are usually divided into three parts: (1) general clauses and agreements, (2) technical instructions, and (3) acceptance of the system.

General clauses and agreements (sometimes called "boilerplates") can also contain descriptions of a quality assurance program; shipping, handling, and storage of materials; equipment delivery; administrative procedures; basis for payment; and project work rules.

Technical instructions affect materials and workmanship, and contain the scope of the project, applicable referenced documents, design and fabrication requirements, erection/installation requirements (i.e., workmanship), and design documents.

Specifications should contain all the necessary instructions and requirements regarding equipment, performance of the system, materials used, and workmanship. Furthermore, they must be clear, concise, correct and unambiguous. (Contractors have included extra charges for projects with poor specifications.) An additional part of the specification is "acceptance of the system," which requires an acceptance test of the system and subsequent certification of the system. (See Chapter 14, "Testing.")

Definitions

Typical standard clauses in specifications include such definitions as:

Engineer. The chief engineer, acting either personally or through duly authorized representatives, who operates within the scope of the authority vested in him/her.

Contract. The entire agreement between the parties, including the form of contract together with the specifications, the advertisement, the information for bidders, the contractor's proposal (copies of which are bound with or accompany the other data), and the contract drawings.

Owner. The corporation.

Contractor. The individual, corporation, company, firm, or other organization who or which has contracted to furnish the materials and to perform the work under the contract.

Work. All labor, plant, materials, facilities, and any other items necessary or proper for or incidental to the construction of the fire alarm signaling system.

Contract Drawings. Those listed in specifications under the title "contract drawings," and any future changes and revisions of said drawings.

Inspector. Any representative designated by the engineer to act as an inspector within the scope of the engineer's authority, as delegated. The engineer, however, may review and change any decision of an inspector.

Subcontractor. Anyone other than the contractor who performs work at the construction site directly or indirectly for or in behalf of the contractor. "Subcontractor" does not include someone who furnishes personal services only.

Writing Specifications

Specifications are clear, concise, and appropriate if they are well organized, well written, and follow proper rules of grammar. The following should be included:

1. A table of contents with page references and an index should be prepared.
2. A numbering system for the various sections and clauses in the documents should be established for easy future reference.
3. All parts to be covered should be outlined before the clauses are prepared. The specification document should be divided into appropriate sections.
4. All hazards and contingent information should be revealed.
5. Requirements in the specifications should be fair, reasonable, and practicable.
6. A given point should be researched if the subject matter is ambiguous or incomplete.
7. Cross references with title and paragraph numbers should be clearly noted.
8. The written material should be coordinated with the work of persons who are making the drawings. The drafter or person who prepares documents should examine all related written materials.

9. Wording should be clear and precise. Objective review is encouraged.
10. Proper rules of grammar and punctuation must be followed. Be concise: short sentences are preferable over long, involved ones.
11. A specification is a legal document in which accuracy and clarity are critical. Complete sentences, not phrases, must be used.
12. Use of ambiguous words, words with more than one meaning, and colloquialisms should be avoided. Avoid synonyms and write in simple, clearly understood language.
13. Relative and personal pronouns, such as "his" and "it," should be avoided.
14. All implications of broad, general statements should be examined before such statements are used.
15. Omissions and irrelevant or useless repetitious information should be avoided.

There are two general types of specifications used for fire alarm signaling systems: (1) vendor's specifications that concern equipment and (2) engineering specifications that concern system performance.

Key parameters in fire alarm signaling system specifications should define and indicate the following requirements:

1. Type of system based on applicable codes.
2. Fire alarm control unit(s) location.
3. Primary (main) and secondary (standby) power supplies required.
4. Interfaces and responsibilities for:

    (a) waterflow switches,
    (b) fire suppression system(s),
    (c) existing facility's electrical system, and
    (d) fire department alarms.

5. Location of fire detectors and alarm boxes. These locations should be dimensioned on drawings prepared by the contractor.
6. Listed equipment.
7. Bill of materials.

Example Specification

The following example specification illustrates typical components of an invitation to bid. It shows information that might be needed for a fire alarm signaling system but is not intended to be a model specification.

TECHNICAL SPECIFICATION FOR FIRE DETECTION AND ALARM SIGNALING SYSTEMS (EXAMPLE)

1.0 *Purpose*

Fire detection and alarm signaling systems are critical to ensuring life safety. They also protect valuable property, so it is imperative that these systems function properly and reliably. This specification has been prepared to ensure that all installations of these systems work properly.

2.0 *Applicability*

These requirements are applicable to new installations of fire detection and alarm signaling systems in buildings owned or leased by (company name). (These requirements are also applicable to modifications in existing systems.)

3.0 *Revisions and Circulation of Specification*

The owner is responsible for periodic review and revision of this specification. The (owner's name) may be reached at (owner's address, responsible person, and phone number).

Revisions become effective on the date of publication unless otherwise indicated. Contracts for fire detection and alarm systems are governed by the issue in force on the bid opening date.

Copies of this specification will automatically be provided to the Division of Purchase and Contract, designers of alarm systems, equipment manufacturers, and other interested parties.

4.0 *General Requirements for All Systems*

Approval of samples, cut sheets, shop drawings, and other matter submitted by the contractor must not relieve the contractor of responsibility for full compliance with the specifications, unless the attention of the engineer is called to each noncomplying feature by letter accompanying the submitted matter.

4.1 All fire alarm signaling systems must comply with (add appropriate codes), unless otherwise approved by the company.

4.2 The system must be nominal 24 volts dc, and noncoded. All equipment supplied must be listed for the purpose for which it is used, and installed in accordance with any instructions included in its listing. Equipment must be new, with a minimum one-year warranty on parts and labor from the date of final inspection and acceptance by the owner.

4.3 The fire alarm control (FAC) panel must be of modular design, for ease of future system extension or modification, or both. The front of the panel must have a steady "power on" light (green color), and each zone module must have separate "alarm" (red color) and "trouble" (amber color) indicators. Zones should be configured in accordance with the contract drawings. (See Table 11-1.)

TABLE 11-1. Zoning Fire Alarm Systems

Alarm Initiating Device Circuits:
   1. Minimum of one zone for each building floor.
   2. Each zone not to exceed 20,000 square feet (1858 m²) or 200 (61 m) linear feet.
   3. Establish zone boundaries at smoke/fire partitions (if any) and/or other readily identifiable building features.
   4. For high-rise buildings that exceed 75 ft (23 m) in height, separate zones for smoke, manual, and waterflow alarms on each floor.

Alarm Indicating Appliance Circuits:
   1. Maximum coverage is one floor, with total load not to exceed 75% of rated module output.

4.4 The FAC power supply must have a continuous rating adequate to power all zones and functions indefinitely in full alarm condition.

4.5 The system must be equipped with protective devices to prevent damage or nuisance alarms by nearby lightning strikes, stray currents, or line voltage fluctuations.

4.6 Both audible and visible alarm signals must be provided. Visible signals must be the strobe (flash discharge) type, with a minimum intensity of 8,000 candlepower. When installed in corridors, visible signals must be equipped with a side-viewing lens.

4.7 The FAC panel must have an "alarm silence" switch. Subsequent alarm (alarm resound) feature is required in all buildings with two or more zones. Any annunciators/graphic displays remote from the area alarmed must also include an audible signal with subsequent alarm.

NOTE: Annunciation and graphic display requirements should be defined in this section of the specification.

4.8 The coverage of each fire alarm zone must be clearly indicated on each FAC panel and remote annunciator. This indication may be by engraved labels, framed directories, graphic display, or any combination. Use of label tape or handwritten labels for this purpose is not acceptable.

4.9 Detectors used for elevator capture must be on a separate (vertical) zone. The capture signal must come from the FAC panel, and be triggered by elevator lobby smoke only. Detector-based relays must not provide this signal but may be used to report the floor that has elevator lobby smoke by initiating an alarm in the adjacent corridor zone.

The operation of automatic fire extinguishing systems should also be separately annunciated, since this normally means that a substantial fire is in progress.

4.10 Systems are to be provided with a separate and independent source of emergency standby power. Switching to emergency standby power during alarm must not cause signal drop-out. Batteries must meet the appropriate capacity requirements of the local code plus a 25 percent safety factor.

If NFPA standards are followed, limited standby battery capacity for the alarm system is still required when a single generator is used. Further information can be found in NFPA 72, *National Fire Alarm Code.*

4.11 A proprietary multiplex system must have Style "_____" initiating device circuits, and Style "_____" signaling line circuits in compliance with NFPA 72, *National Fire Alarm Code.* (Appropriate style letters should be added in this section.)

4.12 All wiring must be appropriately color coded, and permanent wire markers must be used to identify the terminations for each zone at the control panel. The number of splices must be held to an absolute minimum. All junction boxes that are visible or accessible must be painted red, unless in finished areas.

NOTE: The engineer can specify wire markers at all junction boxes, instead of color-coded wiring, if necessary on large/complex systems. However, both color coding and wire markers are desired.

4.13 Wiring must be in metal conduit or surface metal raceway, and copper conductors must be used. The connectors should be sized in accordance with the equipment manufacturer's recommendations but in no case are alarm initiating circuits to be less than 16

AWG stranded/14 AWG solid, or alarm indicating circuits less than 14 AWG.

Stranded wire must comply with Section 760-16(a) of NFPA 70, *National Electrical Code*, which requires that 16 AWG stranded wire be bunchtinned (bonded) or have 7 strands maximum.

4.14 Initiating device or indicating appliance circuits must not be included in raceways containing ac power or ac control wiring. Within the FAC panel, any ac control wiring must be properly separated from other circuits. The enclosure must have an appropriate warning label to alert service personnel to the hazard.

4.15 To prevent insulation damage, the following requirements apply:

Any threaded rigid conduit terminating at sheet metal boxes and cabinets must be provided with insulating bushings.

Any electrical metallic tubing (EMT) connectors must be the all-steel compressing type, with insulated throats. Exception: The indenture type can be used if run exposed in finished areas, to obtain a tighter fit to the surface.

Any surface metal raceway ("wiremold") must be carefully reamed to remove all burrs and sharp edges. Plastic bushings must be used at all raceway terminations.

4.16 All wiring must be checked for grounds, opens, and shorts prior to termination at panels and installation of detectors. The minimum resistance to ground or between any two conductors must be 10 megohms, verified with a megger.

NOTE: The owner must be given advance notice of these tests, so the owner or a representative can be present at the tests.

4.17 The system must be electrically supervised for open or $(+/-)$ ground fault conditions in the alarm initiating circuits, the alarm indicating circuits, and the system alarm and trouble relay coils. Removal of any detection device, alarm appliance, system module, or standby battery connection must result in a trouble signal. A fire alarm signal must override trouble signals, but any pre-alarm trouble signal must reappear when the panel is reset.

4.18 If the FAC panel is in an unoccupied area or area not normally occupied, a remote annunciator, with an audible-visible trouble signal must be provided. The FAC panel area must be protected by a minimum of one smoke detector within 15 linear feet (4.5 m) of panel location.

4.19 Manual fire alarm boxes must require the use of a key or Allen-type wrench for reset.

If the specification concerns a school, dormitory, or other place with many occupants, the following note should also be included:

NOTE: False alarms at manual fire alarm boxes can be a serious problem in schools, dormitories, and other occupancies. In such cases, each manual box should be covered with a metal box that has a glass panel that can be broken to gain access. The optional key lock feature is recommended for convenient test or glass replacement.

5.0 *Requirements for Detectors*

5.1 Detectors must be the plug-in type, with a separate base to facilitate replacement and maintenance. Each detector or detector base must incorporate a light to indicate if it is in alarm.

5.2 All air duct/plenum detectors and underfloor detectors must have a remote alarm indicator light, located in the nearest corridor or public area, and identified with a permanent label affixed to the wall or ceiling. Air duct/plenum detectors and underfloor detectors must also be installed in a manner that provides suitable access for periodic cleaning and calibration.

NOTE: Interlocking acoustical ceiling tile does not provide "suitable access." For more information, refer to NFPA 90A, *Standard for the Installation of Air Conditioning and Ventilating Systems.*

5.3 Open-area-type detectors mounted within 12 ft (3.7 m) of a walking surface must have their built-in locking device activated to prevent unauthorized removal.

5.4 HVAC ducts, lighting fixtures, and other factors affecting air flow must be considered in establishing the exact location of the detectors.

5.5 Open-area-type detectors should not be within 3 ft (0.9 m) of an air supply.

5.6 Unless suitably protected against dust, paint, and other material, detectors must not be installed until the final construction cleanup has been accomplished.

NOTE: A detector that is surface mounted rather than flush mounted responds better to smoke. Therefore, surface mounting must be used unless the detector has been listed for flush mounting.

5.7 CAUTION: Smoke detectors vary widely in their electrical characteristics and must be carefully matched with a suitable control panel to ensure proper performance. The contractor must verify and certify the compatibility of the detectors to the owner's satisfaction. (See Chapter 4, "Signal Transmission," for further information on compatibility.)

5.8 CAUTION: The limited power available from an individual smoke zone module generally cannot be relied upon to alarm more than one to three detectors in the zone, depending on whether they have remote alarm lamps and/or auxiliary control relays. HVAC shutdown and other control functions must be accomplished from the FAC panel rather than by individual smoke detector relays.

5.9 CAUTION: Duct smoke detector mounting position and air sampling tube orientation are critical for proper detector operation. The equipment manufacturer's detailed installation instructions must be followed.

5.10 A unique identification number must be assigned to each detector. (Identification should be by zone number and device number within the zone.) This number must be noted on the plans, and also be permanently mounted adjacent to the detector or affixed to its base.

5.11 Detector trouble contacts (if any) must be series-wired electrically between the last alarm initiating device and the end-of-line device.

5.12 The location of detectors must be dimensioned on the plans.

5.13 The temperature rating of the heat detectors should be indicated on the plans and bill of materials.

6.0 *Motorized Fire Dampers (Smoke Dampers)*

6.1 The interface between fire/smoke dampers and the fire alarm signaling system should be indicated here.

7.0 *Automatic Smoke Doors/Automatic Door Locks*

7.1 Automatic smoke doors must either have wall-mounted magnetic door holders and separate heavy-duty closers, or combination door control/smoke detector units.

8.0 *Sprinkler System Connections*

8.1 Installation of sprinkler systems must comply with NFPA 13, *Standard for the Installation of Sprinkler Systems.*

8.2 Waterflow switches must be listed for the purpose.

8.3 Each waterflow switch must be provided with an integral 15-sec time delay device to prevent nuisance alarms from surges in water pressure.

8.4 Permanent provision must be made for testing each switch by waterflow equivalent to that from a standard ½-in. (12.7-mm) sprinkler.

8.5 Sprinkler gate valve supervision must not utilize the waterflow alarm circuit to indicate "trouble" on the waterflow alarm circuit.

9.0 *Verification of System Performance (Acceptance Test)*

9.1 Upon completion of the installation, the contractor and the manufacturer's authorized representative together must test every alarm initiating device for proper response and zone indication, every alarm signaling appliance for effectiveness, and all auxiliary functions such as capture of elevators and control of smoke doors/ dampers, HVAC systems, and pressurization fans. This may be referred to as the "100 percent system test." Personnel (other than the contractor) who installed or furnished any device or appliance should be notified of the test and asked to supply appropriate information.

The owner and a designated representative must be given the opportunity to witness these tests. An itemized test report must then be submitted to the owner, detailing and certifying all results, including the measured sensitivity of each smoke detector. The data for each smoke detector must include the manufacturer's serial number (if any), plus specific location information adequate to quickly pinpoint the device. (See Section 5.10 of this specification.)

After completion of the 100 percent system test and submission of the certified test report, the contractor must set up a "pre-final" inspection with the owner or a designated representative. The system must operate for at least 30 days prior to this inspection.

9.2 The system must be inspected and functionally tested by the owner or designated representative, on a sample basis.

9.3 The warranty period must begin after successful completion of the owner's inspections and tests. In the event of any system malfunctions or nuisance alarms, the contractor must take appropriate corrective action. This action can necessitate a repeat of the 100 percent system test if the owner so desires. Continued improper performance during warranty is cause to require the contractor to remove the system.

10.0 *System Documentation, Operator Training, and Maintenance*

10.1 The contractor must provide the owner with three copies of the following:

1. As-built wiring and conduit layout diagrams, incorporating wire color code and/or label numbers, and showing all interconnections in the system.

2. As-built schematic wiring diagrams of all control panels, modules, annunciators, and communications panels.

3. As-built fire detector location drawings showing location dimensions of each detector and alarm box.

4. Copies of the manufacturer's technical literature on all major parts of the system, including detectors, manual stations, signaling appliances, alarm panels, and power supplies.

10.2 The manufacturer's authorized representative must instruct the owner's designated employees in proper operation of the system and all required periodic maintenance. This instruction must include two copies of a written summary in booklet or binder form that the employees can retain for future reference. Basic operating instructions for the system must be suitably framed and mounted at the FAC panel.

10.3 The contractor must have the manufacturer's authorized representative provide a quote to the owner for regular preventive maintenance. This quote provides the owner with information on internal versus contract cost for the maintenance. This quote must include the following services, after expiration of the standard warranty:

Maintenance every 2 months:

1. Test all waterflow alarm switches.

Maintenance every 6 months:

1. Inspect all FAC panels.
2. Operate all manual stations.
3. Test all restorable heat detectors functionally.
4. Test all smoke detectors functionally.

Maintenance every 12 months:

1. Clean all smoke detectors, including duct types.
2. Measure, record, and set sensitivity of each smoke detector.
3. Verify all auxiliary system functions.
4. Test standby batteries.
5. Verify operation of all audible and visible alarm appliances.

BIBLIOGRAPHY

NFPA Codes, Standards, Recommended Practices, and Manuals. (See the latest *NFPA Catalog* for availability of current editions of the following documents.)

NFPA 13, *Standard for the Installation of Sprinkler Systems.*

NFPA 70, *National Electrical Code.*

NFPA 72, *National Fire Alarm Code.*

NFPA 90A, *Standard for the Installation of Air Conditioning and Ventilating Systems.*

NFPA 170, *Standard for Fire Safety Symbols.*

12

Approvals and Acceptance

INTRODUCTION

The fire alarm signaling system designer is responsible for obtaining proper approvals in many phases of system design and subsequent installation, acceptance testing, and periodic test and maintenance. This chapter describes the function of the authority having jurisdiction (AHJ) concerning a fire alarm system, types of standards and the standards development process, and listing tests for fire alarm system components. For related information see Chapter 13, "Installation," and Chapter 14, "Testing."

APPROVALS, AUTHORITIES, LABELING, AND LISTING

Standards for fire alarm signaling systems, components, and related equipment contain the terms "approved," "authority having jurisdiction," "labeled," and "listed." The definitions of these terms are important to the fire alarm signaling system designer because many applications of fire alarm signaling systems comply with requirements of public and private agencies. System manufacturers, as well, build systems using these requirements. Generally, the terms can be defined as follows.

"Approved" means acceptable to the authority having jurisdiction. Authority having jurisdiction is the organization, office, or individual responsible for approving equipment, an installation, or a procedure.

"Labeling" is defined as a label, symbol, or other identifying mark that has been attached to equipment or materials by an organization that is acceptable to the authority having jurisdiction.

"Listing" refers to the equipment and materials that are included in a list that is acceptable to the authority having jurisdiction.

The labeling or listing organization is concerned with product evaluation, and conducts periodic inspections of production of labeled or listed materials and equipment. A manufacturer with labeled equipment is indicating compliance with appropriate standards or specified performance; manufacturers with listed equipment are stating either that the equipment or material meets appropriate standards, or it has been tested and found suitable for use in a specified manner.

In the approval process, the authority having jurisdiction may base acceptance of a fire alarm signaling system installation, procedures, equipment, or materials on compliance with appropriate standards. Instead of using standards, the authority may require evidence of proper installation, procedure or use, or refer to listing or labeling practices of a product evaluation organization. This organization will determine compliance of the system with appropriate standards for the current production of listed items.

The definition of "authority having jurisdiction" can vary. In NFPA documents, the authority is defined broadly, since jurisdictions and approval agencies vary depending on their responsibilities. Where public safety is the paramount consideration, the authority may be a federal, state, local, or other regional department, or an individual such as a fire chief, fire marshal, chief of a fire prevention bureau, the head of the labor or health department, a building official, electrical inspector, or other individual with statutory authority. The authority, for insurance purposes, could be an insurance inspection department rating bureau or other insurance company representative. At a government or military installation, the authority could be the command officer or departmental official.

"Labeling" and "listing" are not necessarily interchangeable terms, since some organizations do not recognize equipment as listed unless it is also labeled. A listed product should be identified by the system employed by the listing organization. Most authorities having jurisdiction rely on the labeling or listing of organizations that it finds acceptable.

In many projects, a system designer is responsible for specifying the installation of a system and, unless he/she has delegated this responsibility, the designer in a sense is an authority having jurisdication for all tradespersons responsible for implementation of the system design. Depending on terms of the contract with the facility owner, the designer may be directly responsible for system compliance with the requirements of the authority, or may report to an owner. The owner of the protected facility is ultimately responsible for a system being acceptable to a local authority, although requests for approval at certain stages of a project are usually handled by the individual trades involved.

It is important for the designer to know the identity of the authority since a request for approval at all stages of a project must be submitted to the proper organization or individual. Proceeding without approvals may cause work to be redone at later stages in the project, and at greater expense.

Requests for an authority's approval in a project may take various forms. In general, standards requiring approval also require sufficient data to be submitted at the time the approval is requested so the authority can base approval on compliance with job requirements. The submission of data can be a simple certification that a system has been installed in a given building to a given standard or code. The submittal can also be more complex, with detailed data on equipment specifications, wiring diagrams, floor layouts, description of operation, dimensions, etc. Complex submittals may have to be coordinated depending on the levels of project management and the detail of reporting required by project managers.

CODES AND STANDARDS

Codes and standards fall into three broad categories: (1) codes, (2) installation standards, and (3) performance standards.

Codes specify circumstances under which a given type of protection is required. For example, NFPA *101*, *Life Safety Code*, requires specific types of systems and devices for specific occupancies. For instance, NFPA *101* requires a corridor smoke detection system and one or more manual fire alarm stations for new hotels, depending on whether the hotel is sprinklered. An annunciator panel and an emergency voice/ alarm communication system is also required, depending on building size and height.

Installation standards detail how the protection specified by the code is to be achieved. In addition to details on installation and use, these standards include requirements for maintenance and periodic testing of the installed equipment.

Performance standards specify which functions and capabilities are required of the hardware and conditions under which the equipment must operate. Standards developed by testing laboratories, such as Underwriters Laboratories Inc. (UL) and Factory Mutual Research Corporation (FMRC), are examples of performance standards. For the hotel example previously cited, UL 268, *Standard on Smoke Detectors*, describes tests for listing the smoke detectors used in hotel corridors. The standard includes requirements for basic sensitivity and the maximum acceptable change in sensitivity allowed when the detector is operated under various environmental conditions.

Codes can also contain installation and performance requirements, while installation standards can also include performance details. NFPA 101, *Life Safety Code*, Section 7-6, contains installation and performance-type requirements for fire detection and alarm systems; NFPA 72, *National Fire Alarm Code*, contains performance requirements for specific devices. Requirements repeated in two or more standards usually do not conflict; but if conflicts do occur, the designer should propose a solution that is acceptable to the authority having jurisdiction.

The Standards Development Process

Most standards used in the United States are developed under a consensus process. Initially, a sponsoring organization develops a draft standard. In some cases, public notice of the intent to develop such a standard is given, and proposals concerning the standard are solicited from interested parties, e.g., manufacturers, building officials, government officials, private citizens, fire service representatives, etc.

This draft standard is then circulated to a broad group of interested parties for their review and comments. Comments on the draft are reviewed and considered by the originating body and are accepted, negotiated, or rejected. When the draft comments are negotiated or rejected, this process is carefully documented so the reasons for the action are clear.

The next step in some standards-making organizations is a formal voting procedure. For the National Fire Protection Association (NFPA), the Association's members who attend one of the NFPA's two member meetings each year will vote to accept or reject the draft. Approval laboratories without a voting membership often submit their standards to a review and voting cycle administered by the American National Standards Institute (ANSI). ANSI recognition of a standard represents the completion of the consensus process.

Following completion of the consensus process, these standards are adopted by federal, state, or local government as part of their building or fire codes and become law. The standards may be adopted completely, in part, or with modifications. Code requirements for any system can vary from one jurisdiction to another, depending on how they were adopted.

Final determination of how a standard is applied rests with the person or organization referred to in the codes as the authority having jurisdiction. (NFPA standards define the authority as the organization, office, or individual responsible for approving the equipment and installation, or procedure.) Specific aspects of system design and operation may need to be approved by the authority before it is considered that the requirements of the standard have been satisfied.

Equivalency

Most standards in the United States contain an "equivalency" clause, which allows for the use of systems, methods, or devices of equivalent or superior quality, effectiveness, or safety that are different from those prescribed by the standard, as long as sufficient documentation is provided to demonstrate the equivalency to the satisfaction of the authority having jurisdiction.

Equivalency is critical to the fire alarm signaling system designer. Codes and standards are written in general terms and represent only the minimum acceptable requirements for general applications of systems and equipment. The designer should design a system to meet the intent and goals of safety for the specific application; but, in some circumstances, particular conditions or alternate methods are not implied by the code, and equivalent measures can be justified. A designer should use good engineering judgment when applying for acceptance under the equivalency clause.

There are no established rules, as of this writing, for demonstrating equivalency. The performance of, and reference to, fire tests or other available experimental data has been used to show equivalency. However, computerized building fire simulation modeling and calculating methods are being used increasingly and allow for easier demonstration of equivalency.

International commerce and the standards foreign countries require U.S. manufacturers to meet are more complex than U.S. requirements. Foreign product approval systems tend to also use a consensus process, but it can differ from one country to another. In most Western European countries, the process is developed by manufacturing, insurance, and government organizations and is similar to the U.S. system; in Japan, the system is administered by the government and is mandatory.

A complicating factor in international product approval is that a large number of different approvals are often required, and may include one approval per country or group of countries in which the product will be sold. This situation led to the creation of the International Standards Organization (ISO), which is working to correlate different standards and develop consistent international requirements. The activities of ISO Technical Committee 21, which develops test standards for fire detectors and fire alarm systems, is expected to result in a common set of fire detection and alarm standards. Subsequently, approval tests in international markets could be conducted in a single laboratory and a test report then submitted to other testing laboratories from which approval is desired. Any individual approval laboratory would be free to establish its own requirements for regular production testing and quality assurance.

TESTING LABORATORY PROCEDURES

In the United States a testing laboratory develops a product safety standard and tests samples of the product to that standard. A fire alarm signaling system designer should be familiar with the detailed inspection and quality assurance procedures that comply with a testing laboratory listing.

Once samples of the original product have passed all the tests and been listed, inspection procedures for the product are developed by the engineer who conducted the original investigation for the laboratory. Periodic unannounced plant inspections are conducted on production units to ensure that no changes are made that might result in unacceptable performance. These inspections include comparisons of the product being assembled to detailed descriptions and drawings of the originally tested item, and also often involve a detailed evaluation of the manufacturer's quality assurance program and verification process. This inspection system mandates that any and all changes to the product be submitted to and approved by the testing laboratory before such changes are incorporated into production units. By contractual agreement, UL and FMRC, two of the major testing laboratories in the United States, prevent shipment of a product bearing a mark that may not meet the requirements of their individual standards.

Heat Detector Testing

UL, in testing to list heat detectors, determines heat detector ratings as "spacings." Spacing of 15 to 60 ft (4.5 to 18 m) are given based on actual tests. (See Figure 12-1.)

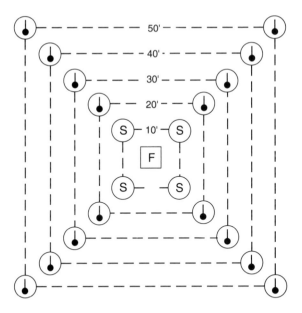

LEGEND

F – TEST FIRE, DENATURED ALCOHOL, 190 PROOF. PAN LOCATED
      APPROXIMATELY 3 FT (0.9 M)ABOVE FLOOR.

(S) – INDICATES NORMAL SPRINKLER SPACINGS ON 10-FT (3-M)
      SCHEDULES.

(⬥) – INDICATES NORMAL DETECTOR SPACING ON VARIOUS SPACING
      SCHEDULES.

FOR SI UNITS: 1 FT = 0.30 M.

Figure 12-1.   Layout for UL heat detector test.

Heat detectors are tested by UL with an alcohol pan fire. In each of two tests, the time of operation is measured for 160°F (71°C)-rated automatic sprinklers that are installed on a 10 × 10 ft (3 × 3 m) spacing. When a fire of sufficient size is reached to operate the sprinklers in 2 min after ignition, the heat detector installed on its maximum spacing must have operated *before* the automatic sprinkler.

Operation of the sprinkler head is the point of comparison and the time-temperature curve is similar to the 15-ft (4.6-m) spacing curve shown in Figure 12-2. If earlier warning is desired, less than listed spacing will probably be necessary.

The distance that heat detectors are located from the fire is representative of the spacing for which the detector is being tested. This distance is also the maximum distance from the fire allowed for proper installation (which will also result in the slowest acceptable detector response).

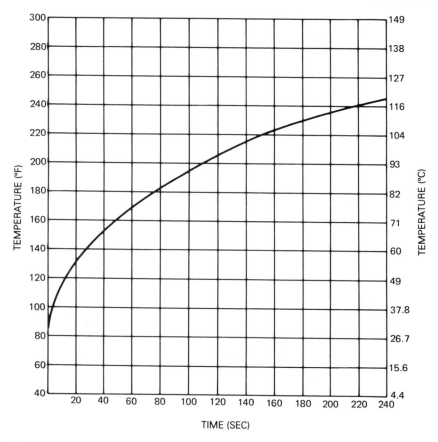

Figure 12-2. Heat detector time-temperature curve — 15-ft (4.6-m) spacings.

The spacing curve shows an approximate air temperature at the detector of 206°F (97°C) when detection occurs. (See Figure 12-2.) This illustrates the effect of thermal lag, which is the difference between the set point of the detector [perhaps as low as 135°F (57°C)] and the air temperature required for the detector to operate in a fire condition.

Smoke Detector Testing

A smoke detector is a device that detects visible or invisible particles of combustion.

Numerous field tests have shown that heat detectors do not provide the speed of response to a fire needed for life safety. Two of the major field test programs were conducted during the 1970s in Indiana. The conclusions of the second test series state, "Supporting the first year

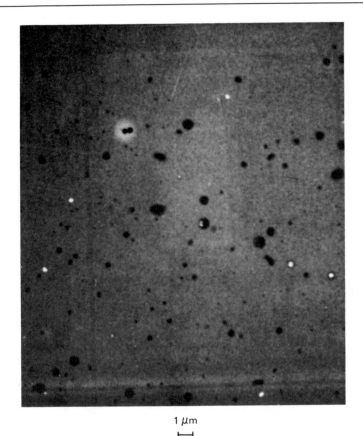

1 μm
⊢─┤

Figure 12-3.   Fresh Douglas fir smoke particles.

results, the fixed temperature heat detectors rated for 50-ft (15-m) spacing [135°F (57°C)] used in this test series, in the room of fire origin, provided little life saving potential. These detectors failed to respond to a majority of the fires and when they did respond they were considerably slower than smoke detectors located remotely from the fire."[1]

Since smoke detectors do respond much faster to a fire than a heat detector, the test fires they are required to detect are much smaller than the test fires used for the listing of heat detectors. Also, since smoke detectors operating on different principles of operation respond differently to smoke from different combustibles, UL has established five test fires, each with a different combustible, that every detector must detect.

A further need for different small test fires for smoke detectors is summarized as follows: "The essential feature of smoke is its instability. Under the influence of a lively Brownian motion, the particles collide

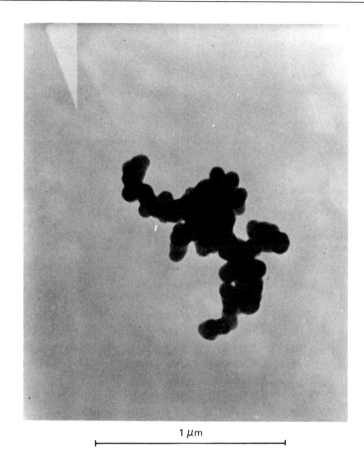

1 μm

Figure 12-4. Aged Douglas fir smoke particles.

with one another and agglomerate. This process goes on continuously until the number of particles has been considerably diminished and the average size largely increased."[2] This effect can be seen in Figure 12-3 and Figure 12-4, which are photos of fresh Douglas fir smoke particles and the same smoke after it has aged for a period of time, respectively.

Prior to conducting fire tests, smoke detectors are placed in a smoke box similar to the one shown in Figure 12-5. The detector is mounted in a horizontal position at the top of the upper chamber; smoke is then introduced and circulated across the lower chamber, up through the air straightener to the detector. The smoke density is gradually increased until the detector responds, establishing the sensitivity (set point) of the detector.

Additional tests are conducted with the oncoming smoke directed to each of four sides of the detector and also with the detector placed on

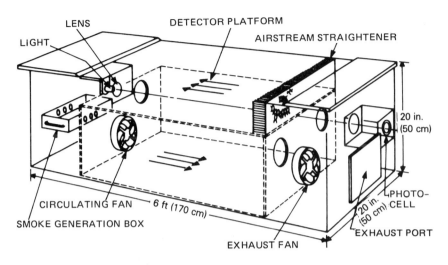

Figure 12-5. Smoke detector evaluation chamber.

end facing the oncoming smoke to determine a least favorable position
of smoke entry (if any). If there is a least favorable entry position, the
detectors are placed in this position with respect to the oncoming
smoke, when mounted in the fire test room. In all cases, the sensitivity
of the detectors should be within the limits shown in Table 12-1.

TABLE 12-1. Visible Smoke Obscuration Limits

A. Visible Smoke Obscuration Limits (Gray Smoke):

Percent Per Foot	*Percent Per Meter*	*OD Per Foot*	*OD Per Meter*	
4.0	12.5	0.0177	0.0581	Maximum
0.2	0.65	0.00087	0.0029	Minimum

B. Visible Smoke Obscuration Limits (Black Smoke):

Percent Per Foot	*Percent Per Meter*	*OD Per Foot*	*OD Per Meter*	
10.0	29.2	0.046	0.150	Maximum
0.5	1.6	0.0022	0.0071	Minimum

C. Measuring Ionization Chamber (MIC) Measurement:

14.5 psi (100 Pa) (minimum) – 5.4 psi (37.5 Pa) (maximum)

OD = optical density.

In addition to the above, the detectors selected for the fire tests are detectors set at the minimum production sensitivity. For the fire tests, all of the detectors except battery-operated single-station detectors are connected to a source of supply at their rated voltage. These detectors are tested with the battery depleted to the trouble signal voltage level.

The five UL test fires to list smoke detectors include two fires that produce a black smoke (combustibles are gasoline and polystyrene), two that produce a gray smoke (wood and newspaper combustibles), and one smoldering fire that produces gray smoke (wood is burned on a hotplate). The first four are flaming fires and the fifth a smoldering fire, so a wide cross section of types of smoke are included in the listing process.

UL uses two methods of smoke measurement during the fire tests. The first method is a light beam with the light source placed 5 ft (1.5 m) from the light-sensitive receiver. (See Figure 12-6.) An instrument measures the light being received (percent of light transmitted) in clear area over this distance and the meter reading is adjusted to register 100 percent. Smoke entering the detector will obscure some of the light and decrease the meter reading. The meter readings for the five fire tests are then plotted on charts as shown in Figures 12-7 and 12-8.

Smoke obscuration of 2 percent per ft (0.3 m) is defined here as smoke that is of sufficient density to obscure 2 percent of the light over a distance of 1 ft (0.3 m). A 5-ft (1.5-m) beam is used instead of a 1-ft (0.3-m) beam to obtain a more dramatic and readable difference at the meter. Since a 5-ft (1.5-m) beam is used, as the light travels through 1 ft (0.3 m) of smoke with a 2 percent per ft (0.3 m) obscuration, it cuts off 2 percent of the light at the 1-ft (0.3-m) position of the beam. During the next 1 ft (0.3 m) of travel, the 2 percent smoke cuts off 2 percent of the light that was left at the 1-ft (0.3-m) mark. As the light continues to travel through the 2 percent smoke, the light transmission is reduced by 2 percent of the light that was at each previous 1-ft (0.3-m) mark. So 2 percent smoke does not cut off 10 percent of the light at the 5-ft (1.5-m) mark, and that is why Figure 12-8 shows that it requires 2.1 percent obscuration per ft (0.3 m) to reduce the percent light transmitted to 90; 4.4 percent for 80; 6.9 percent for 70, etc.

A second method of measuring smoke during test fires is through a measuring ionization chamber (MIC) device, developed by Cerberus Ltd., Switzerland. This method is preferred to test the performance of ionization-chamber smoke detectors.

The MIC characterizes the smoke present in the same reference terms as the ion chamber smoke detector under test. When used in conjunction with a light beam, the mean particle diameter of the generated

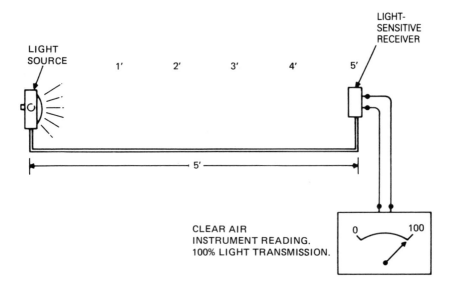

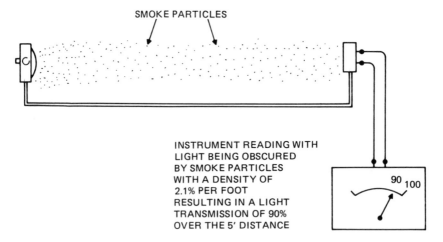

FOR SI UNITS: 1 FT = 0.30 m.

Figure 12-6. Light beam measurement for UL smoke detector test.

smoke can be determined. When a MIC is used with a light beam and a thermocouple, the combination of these three measuring instruments permits the measurement of the three most important parameters of the smoke detector test fires, and can be used to determine whether repeatable fires are being produced (and the particle size of the smoke).

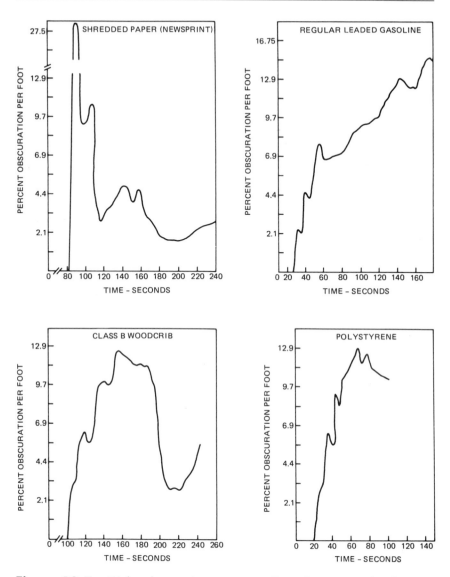

Figure 12-7. Light obscuration meter readings for UL smoke detector tests.

The MIC is designed to be a standard measuring chamber and has a parallel plate electrode configuration in which the alpha source (Americium 241) is part of one of the electrodes. This configuration provides a measuring volume in which the ionization is uniform and approximately parallel to a constant electrical field. (See Figure 12-9.)

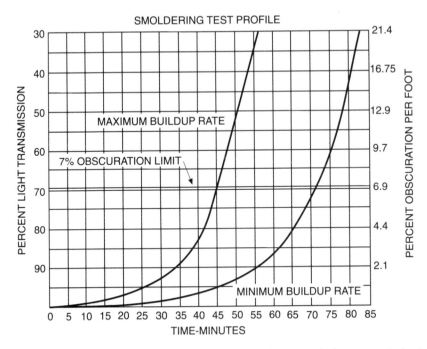

Figure 12-8. Percent obscuration per foot (percent light transmission).

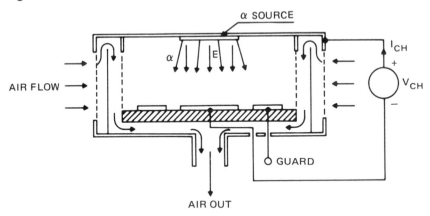

Figure 12-9. MIC configuration.

The air is sucked through the chamber to reduce wind dependence. The air in the measuring volume between the electrodes remains stationary, since the sucked air flows in a duct separated from the measuring volume by wire mesh. Smoke is transferred from the air flow to the measuring volume by diffusion.

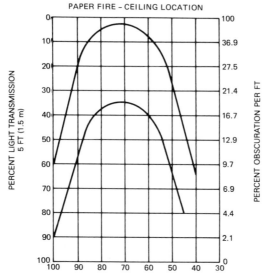

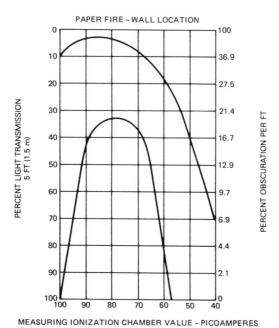

Figure 12-10. MIC profiles for the UL smoke detector test paper fires, ceiling and wall location.

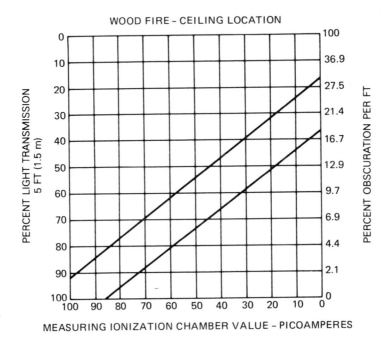

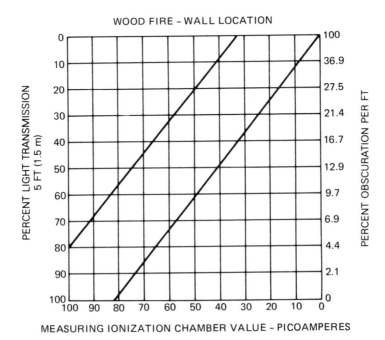

Figure 12-11. MIC profiles for the UL smoke detector test wood fires, ceiling and wall location.

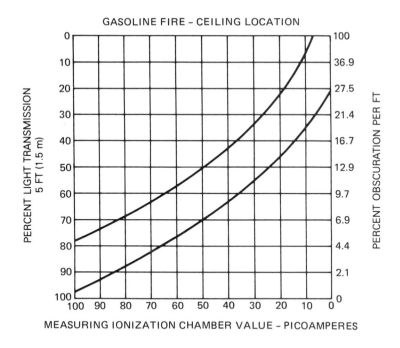

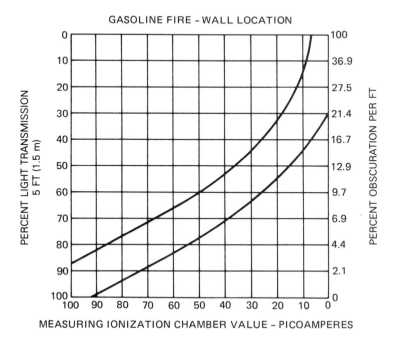

Figure 12-12. MIC profiles for UL smoke detector test gasoline fires, ceiling and wall location.

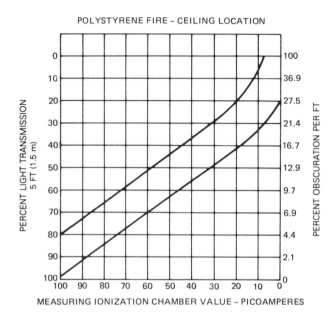

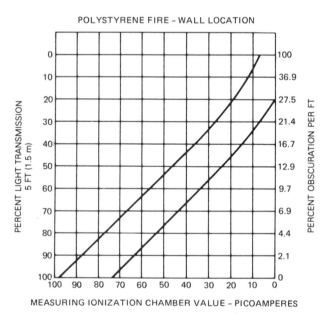

Figure 12-13. MIC profiles for UL smoke detector test polystyrene fires, ceiling and wall location.

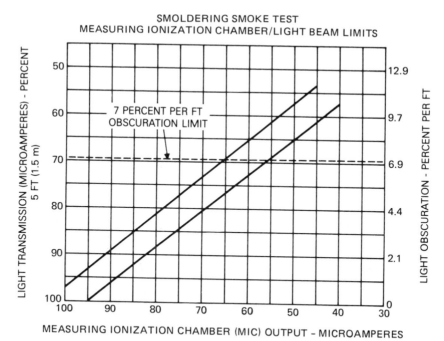

NOTE: LIMITS ARE BASED ON A MEASURING IONIZATION CHAMBER
RESISTANCE OF 20 × 10¹⁰ OHMS MEASURED AT 70° F (21° C), 77 PERCENT
RH, AND 760 mm HG.

Figure 12-14. MIC profile for the UL smoke detector test smoldering fire (measuring ionization chamber/light beam units).

The current of the chamber is measured either directly (with an electrometer) or with a special amplifier. With an amplifier, an impedance transforming circuit is placed inside the MIC, transforming the high impedance level of the ionization chamber to a lower level. This is an advantage when the MIC device is used for a full-scale fire test, because the length of the cable connecting the MIC to the amplifier is not critical. MIC test profiles for all five UL test fires are shown in Figures 12-10 through 12-14.

The five UL smoke detector test fires are conducted in a 36 × 22 × 10 ft (11 × 6.7 × 3 m) test room. (See Figure 12-15.) The test fires are shown in Table 12-2. The detectors are placed on the ceiling and sidewall so they are located 17.7 ft (5.4 m) from the point on the ceiling that is directly over the test fire. Though this spacing does not establish a listed spacing similar to the listed spacing for heat detectors, it does establish a criteria for pass/fail determination of smoke detector response to the smoke created in the five test fires.

TABLE 12-2. Smoke Detector Test Fires (UL)

Test No.	Combustible	Detector Must Respond Within	Typical Smoke Profile	Typical MIC Profile
A.	1.5 oz Shredded Newsprint*	4 min.	Figure 12-7	Figure 12-10
B.	6 × 6 × 2.5in. Layered Fir Woodstrips*	4 min.	Figure 12-7	Figure 12-11
C.	30 ml Regular Leaded Gasoline*	3 min.	Figure 12-7	Figure 12-12
D.	1 oz Foam Polystyrene-Type Packing Material*	2 min.	Figure 12-7	Figure 12-13
E.	Ponderosa Pine Sticks Over Hotplate†	70 min.	Figure 12-8	Figure 12-14

*Flaming Fire
†Smoldering Fire
For SI Units: 1 oz = 28 g; 1 ml = 0.034 oz; 1 in. = 25.4 mm.

Classification (Grading) System for Automatic Fire Detectors

Listing of automatic fire detectors in the U.S. is based primarily on a pass-fail method of fire tests. In some applications, this approach does not provide sufficient data to select the most suitable detector needed. To assist users in detector selection, the European Committee for Standardization (CEN) has proposed that the International Organization for Standards (ISO) consider adoption of a different approach to evaluate fire detectors. CEN test fires consist of a series of test fires using a variety of combustibles as shown in Table 12-3.

The CEN test program classifies each type of fire detector, regardless of operation principle, as a Class A, B, or C, in accordance with its response to each combustible. This classification indicates the sensitivity of a detector type in a particular fire situation. The classification is derived by using three limiting values in each fire test: (1) temperature (T), (2) smoke density [optical (m) with a light beam], and (3) smoke density [ionization (y) with a measuring ionization chamber, or MIC]. (See Table 12-4.)

Four detectors are used in the fire tests and mounted in the fire test room, as shown in Figure 12-16.

If the alarm points of all four detectors are reached before the measuring instruments record ΔT_1, m_1, and y_1, the detector is placed in Class A. If all alarm before ΔT_2, m_2, and y_2, the detector is placed in

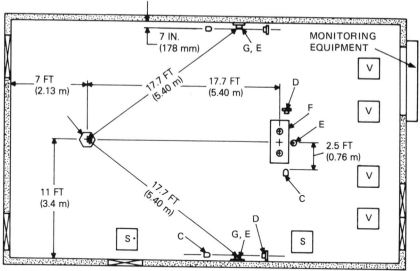

A. FIRE TEST ROOM DIMENSIONS
1. LENGTH - 36 FT (11 m)
2. WIDTH - 22 FT (6.7 m)
3. CEILING - HEIGHT 10 FT (3.0 m) SUSPENDED TYPE. CONSISTS OF 2 BY 4 FT (0.6 BY 1.2 m) BY 5/8 IN. (15.9 mm) THICK NONCOMBUSTIBLE FISSURED MINERAL FIBER LAYER IN PANELS.

B. TEST FIRE
1. 3 FT (0.91 m) ABOVE FLOOR FOR FIRE TESTS
2. 8 IN. (203 mm) ABOVE FLOOR FOR SMOLDERING SMOKE TEST

C. LAMP ASSEMBLY - 4 IN. (102 mm) BELOW CEILING, 7 IN. (178 mm) FROM EACH SIDE WALL.

D. PHOTOCELL ASSEMBLY - SPACED 5 FT (1.5 m) FROM LAMP, PHOTOCELL CENTER 4 IN. (102 mm) BELOW CEILING, 7 IN. (178 mm) FROM EACH SIDE WALL.

E. MEASURING IONIZATION CHAMBER (MIC)

F. TEST PANEL, SIDEWALL-MOUNTED DETECTORS

G. TEST PANEL, SIDEWALL-MOUNTED DETECTORS

S. AIR SUPPLY

V. EXHAUST VENTS

Figure 12-15. UL fire test room for smoke detector test.

TABLE 12-3. CEN Test Fires

Designation (TF = test fire)	Type of Fire	Characteristic Features Development of Heat
TF 1	Open Cellulosic Fire (Wood)	Strong
TF 2	Smoldering Pyrolysis Fire (Wood)	Can Be Neglected
TF 3	Glowing Smoldering Fire (Cotton)	Can Be Neglected
TF 4	Open Plastics Fire (Polyurethane)	Strong
TF 5	Liquid Fire (N-Heptane)	Strong
TF 6	Liquid Fire (Methylated Spirits)	Strong

Class B, and ΔT_3, m_3, for Class C. If the detector does not respond to a specific test fire, it would be classified as N for that combustible.

When the test results are placed on the classification chart, the user has a better picture of detector response to a specific combustible. However, the classification applies only to applications for representative test conditions.

A classification chart for a certain smoke detector might look similar to Table 12-5. Note that since methylated spirits do not produce smoke when burning, the detector did not respond to test fire 6.

Duct Detectors: One method of testing the performance of duct-type smoke detectors is with an air duct facility as shown in Figure 12-17.

The smoke detectors are placed in the test facility at location G. At this point, the duct is 1 sq ft (0.09 m^2) and the detectors are mounted in or on the duct in the same way that they are mounted in the field. The smoke buildup rate, at the various velocities to which they are tested, is shown in Table 12-6. In each test, response to the smoke must be within the limits shown in Table 12-7.

TABLE 12-3. CEN Test Fires (continued)

| Characteristic Features | | | |
Upcurrent	Smoke	Aerosol Spectrum	Visible Portion
Strong	Yes	Predominantly Invisible	Dark
Weak	Yes	Predominantly Visible	Light, High Scattering
Very Weak	Yes	Predominantly Invisible	Light, High Scattering
Strong	Yes	Partially Invisible	Very Dark
Strong	Yes	Predominantly Invisible	Very Dark
Strong	No	None	None

TABLE 12-4. Limiting Values in CEN Classification Program

$$\Delta T_1 = 15°C$$

$$\Delta T_2 = 30°C$$

$$\Delta T_3 = 60\ °C$$

$$m_1 = 0,3\ \frac{dB}{m}$$

$$m_2 = 0,6\ \frac{dB}{m}$$

$$m_3 = 1,2\ \frac{dB}{m}$$

$$y_1 = 1$$

$$y_2 = 2$$

$$y_3 = 4$$

PLAN VIEW OF DETECTORS, FIRE PLACE, AND MEASURING INSTRUMENTS.
THE HEIGHT OF THE ROOM IS 12.5 FT - 14 FT (3.8 m - 4.2 m).
10 FT (3 m)
60°
FIRE PLACE
DETECTORS AND MEASURING INSTRUMENTS
20 FT - 26 FT (6 m - 8 m)
30 FT - 36 FT (9 m - 11 m)

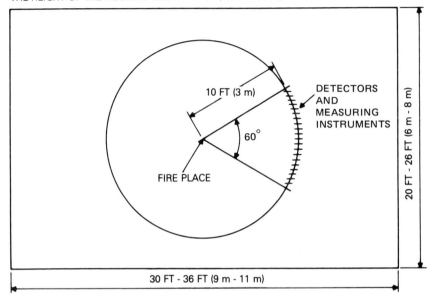

Figure 12-16. CEN/ISO fire test room.

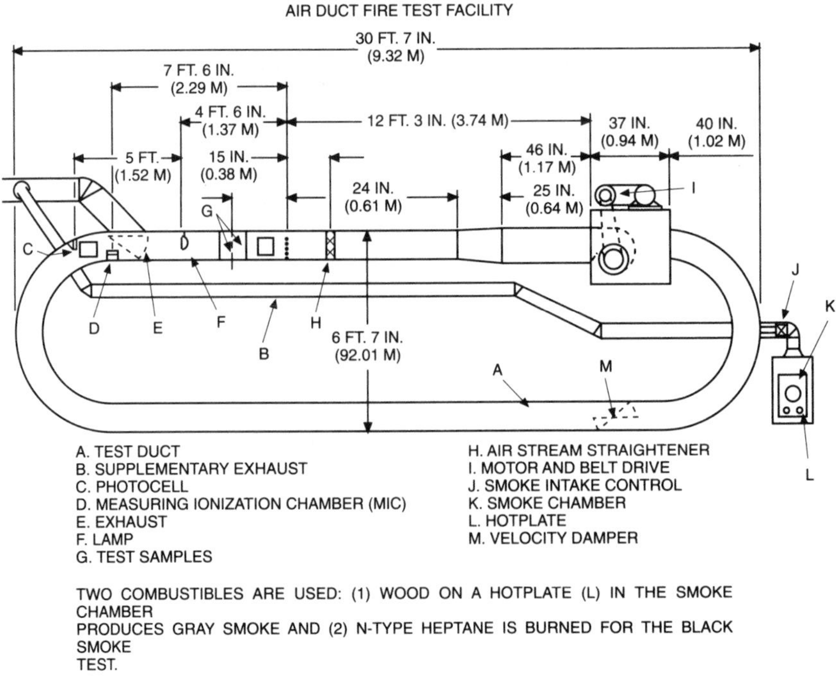

AIR DUCT FIRE TEST FACILITY
30 FT. 7 IN. (9.32 M)
7 FT. 6 IN. (2.29 M)
4 FT. 6 IN. (1.37 M)
12 FT. 3 IN. (3.74 M)
37 IN. (0.94 M)
40 IN. (1.02 M)
5 FT. (1.52 M)
15 IN. (0.38 M)
46 IN. (1.17 M)
24 IN. (0.61 M)
25 IN. (0.64 M)
6 FT. 7 IN. (92.01 M)
A B C D E F G H I J K L M
A. TEST DUCT
B. SUPPLEMENTARY EXHAUST
C. PHOTOCELL
D. MEASURING IONIZATION CHAMBER (MIC)
E. EXHAUST
F. LAMP
G. TEST SAMPLES
H. AIR STREAM STRAIGHTENER
I. MOTOR AND BELT DRIVE
J. SMOKE INTAKE CONTROL
K. SMOKE CHAMBER
L. HOTPLATE
M. VELOCITY DAMPER
TWO COMBUSTIBLES ARE USED: (1) WOOD ON A HOTPLATE (L) IN THE SMOKE CHAMBER
PRODUCES GRAY SMOKE AND (2) N-TYPE HEPTANE IS BURNED FOR THE BLACK SMOKE
TEST.

Figure 12-17. Air duct test facility.

TABLE 12-5. Smoke Detector Classification Chart

Test Fire	Class A	Class B	Class C	N
TF 1	x			
TF 2			x	
TF 3			x	
TF 4	x			
TF 5		x		
TF 6				x

TABLE 12-6. Smoke Parameters

Gray Smoke Parameters

Duct Air Speed FPM	m/sec	Number of Sticks Used in Test	Smoke Buildup Rate, Percent Per Foot Obscuration Per Minute
300	1.52	5	2.9 ± 0.5
1000	5.14	5	2.2 ± 0.8
2000	10.2	6	2.1 ± 0.5
3000	15.4	7	1.3 ± 0 4
4000	20.3	8	0.8 ± 0.2

Black Smoke Parameters

Duct Air Speed FPM	m/sec	Receptacle Size Diameter Inches	(mm)	Depth Inches	(mm)	Amount of Combustible (ml)	Smoke Buildup Rate, Percent Per Foot Obscuration Per Minute
300	1.52	3-3/8	(85.7)	1-7/8	(47.6)	30	4 ± 1.5
1000	5.14	3-3/8	(85.7)	1-7/8	(47.6)	30	4 ± 1.5
2000	10.2	3-3/8	(85.7)	1-7/8	(47.6)	30	4 ± 1.5
3000	15.4	4	(101.6)	2	(50.8)	40	4 ± 1.5
4000	20.3	4	(101.6)	2	(50.8)	40	4 ± 1.5

TABLE 12-7. Visible Smoke Obscuration Limits with Percent Light Transmission

Visible Smoke Obscuration Limits (Gray Smoke)					
Percent Per Ft	Percent Per Meter	OD Per Ft	OD Per Meter	Percent Light Transmission	
4.0	12.56	0.0178	0.0583	81.5	Maximum
0.2	0.66	0.00087	0.0028	99.0	Minimum

Visible Smoke Obscuration Limits (Black Smoke)					
Percent Per Ft	Percent Per Meter	OD Per Ft	OD Per Meter	Percent Light Transmission	
10.0	29.26	0.0458	0.1504	59.0	Maximum
0.5	1.65	0.0022	0.0072	7.5	Minimum

BIBLIOGRAPHY

NFPA Codes, Standards, Recommended Practices, and Manuals. (See the latest *NFPA Catalog* for availability of current editions of the following documents.)
NFPA 72, *National Fire Alarm Code.*
NFPA *101, Life Safety Code.*

References Cited

1. Harpe, S. W., Waterman, T. E., and Christian, W. J., "Detector Sensitivity and Siting Requirements for Dwellings—Phase 2," NBS-GCR77-82, National Institute for Standards and Technology, Gaithersburg, MD, 1976.
2. Hassler, Walter M., "Smoke Detection by Forward Light Scattering," *Fire Technology*, Vol. 1, No. 1, Feb. 1965, p. 43.

Additional Reading

UL 268, *Standard on Smoke Detectors*, Underwriters Laboratories Inc., Northbrook, IL.

Installation

INTRODUCTION

Proper installation of a fire alarm signaling system is obviously a prime factor in subsequent system efficiency and correct operation. Installation also affects any supplemental applications the designer may specify for the system. Listed equipment, adequate power supplies, supervised circuits, trouble signals, and necessary audible alarms should all be considered in the installation process.

This chapter covers installation of automatic fire detectors and special applications of detectors — to control smoke spread, release doors, and actuate various extinguishing systems. Circuit installation and wiring for a fire protective signaling system are also described, and protection of a system from transients and lightning are discussed.

For additional information see Chapter 6, "Electrical Supervision," and Chapter 8, "Power Supplies."

INSTALLATION OF AUTOMATIC FIRE DETECTORS

Detectors should be protected where subject to mechanical damage. They should be supported in all cases independent of their attachment to the circuit conductors. Detectors should not be recessed in any way into the mounting surface unless they have been tested and listed for such recessed mounting.

Detectors should be installed in all areas where required by the appropriate standard in effect for the facility. Where total coverage is required, detectors should be installed in all rooms, halls, storage areas, basements, attics, lofts, spaces above suspended ceilings, other subdivisions and accessible spaces, and inside all closets, elevator shafts, enclosed stairways, dumbwaiter shafts, and chutes. Inaccessible areas that contain combustible material should be made accessible and protected by detector(s).

Detectors can be omitted from combustible blind spaces under any of the following conditions:

1. When the ceiling is attached directly to the underside of the supporting beams of a combustible roof or floor deck.
2. When the concealed space is entirely filled with noncombustible insulation. In solid joist construction, the insulation need only fill the space from the ceiling to the bottom edge of the joist of the roof or floor deck.
3. When there are small concealed spaces over rooms, provided any space in question does not exceed 50 sq ft (4.6 m^2) in area.
4. In spaces formed by sets of facing studs or solid joists in walls, floors or ceilings where the distance between the facing studs or solid joists is less than 6 in. (150 mm).

Similarly, detectors can be omitted from below open-grid ceilings when (1) grid openings are $\frac{1}{4}$ in. (6.4 mm) or larger in the least dimension, (2) the thickness of the material does not exceed the least dimension, and (3) the openings constitute at least 70 percent of the area of the ceiling material.

Detectors can be required under large benches, shelves or tables and inside cupboards or other enclosures. Detectors should also be installed underneath open loading docks or platforms and their covers, and for accessible underfloor spaces of buildings without basements.

Detectors can be omitted when *all* of the following conditions prevail:

1. The space is not accessible for storage purposes or entrance of unauthorized persons and is protected against accumulation of windborne debris; and
2. The space contains no equipment, such as steam pipes, electric wiring, shafting, or conveyors; and
3. The floor over the space is tight; and
4. No flammable liquids are processed, handled, or stored on the above floor.

Connections and Wiring

Duplicate terminals or leads (or the equivalent) should be provided on each automatic fire detector solely to connect into the fire alarm system and provide supervision of the connections. Such terminals or leads are necessary to allow the wire run to be broken and connections to be made to the incoming and outgoing leads or terminals. Detectors that provide the equivalent supervision are excepted. (See Figures 13-1 through 13-5.)

If looped wire connections or unbroken wire runs (secured in a notch) are used and a circuit wire disengages, the detector is rendered inoperative and a trouble signal will not sound. (See Figure 13-2.)

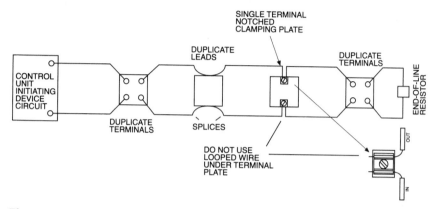

Figure 13-1. Correct wiring method for 2-wire detectors and audible appliances.

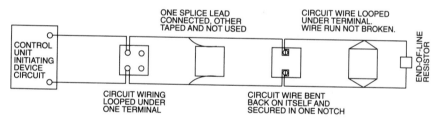

Figure 13-2. Incorrect wiring method for 2-wire detectors and audible appliances.

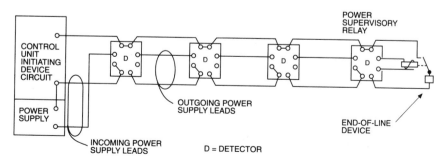

Figure 13-3. Correct wiring method for 3-wire connections.

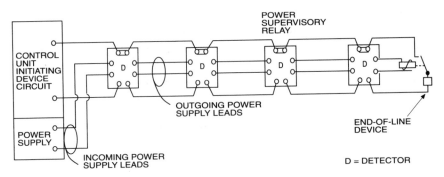

Figure 13-4. Correct wiring method for 4-wire connections.

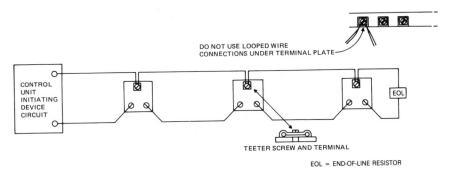

Figure 13-5. Correct wiring method for 2-wire detectors.

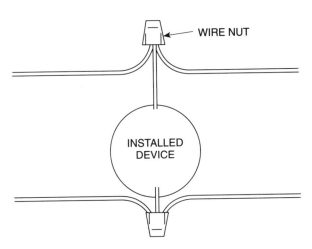

Figure 13-6. Incorrect wiring method for pigtail connections.

Figure 13-6, an incorrect method of wiring, shows that if the conductor from the device to the wire nut becomes disconnected, supervision is still maintained. No indication of trouble would be indicated at the fire alarm panel, even though the device was no longer connected. When the device is wired as shown in Figure 13-7, a trouble signal will sound when any of the conductors is disconnected.

A multiriser initiating or indicating circuit must also be installed correctly so supervision is maintained. (See Figures 13-8 and 13-9.)

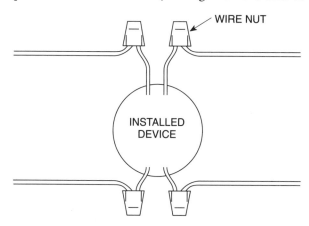

Figure 13-7. Correct wiring method for pigtail connections.

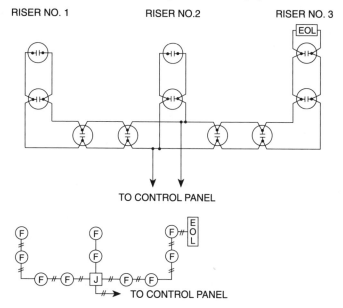

Figure 13-8. Incorrect wiring method for multiriser circuit.

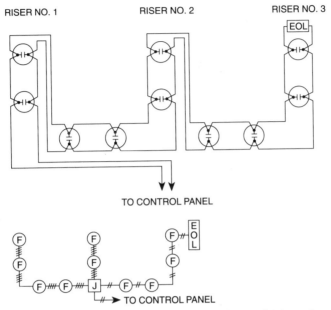

Figure 13-9. Correct wiring method for multiriser circuit.

SPECIAL APPLICATIONS FOR AUTOMATIC FIRE DETECTORS

Fire detectors (smoke detectors) should be properly located and spaced according to the local code. When automatic fire detectors are used in special applications, placement and spacing is often modified from the usual ceiling spacing. This places the detector in the path of the heat or smoke. Detector spacing for special applications is more fully described in NFPA 72, *National Fire Alarm Code*. To minimize dust contamination of smoke detectors when they are installed under raised room floors and similar spaces, the detectors should be mounted only in an orientation for which they have been listed. (See Figure 13-10.)

High Rack Storage

Special care must be taken when installing detection systems in high rack storage areas since the potential fire load makes early detection of fire essential. Detection systems are often installed in addition to suppression systems in high rack storage. Where smoke detectors are installed for early warning in high rack storage areas, the designer should consider installing detectors at several levels in the racks to ensure quicker response to smoke. Detectors can be installed to actuate a suppression system. (NFPA 231C,

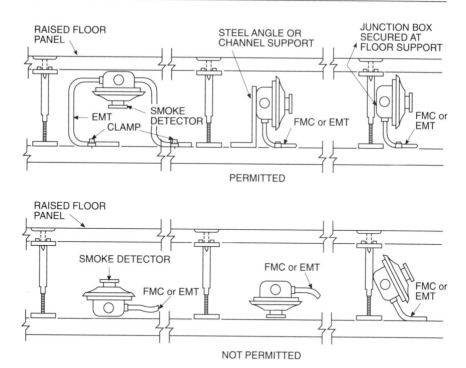

Figure 13-10. Mounting installations for smoke detectors under raised floors.

Standard for Rack Storage of Materials, contains more information on detector actuation of a suppression system in rack storage.)

For the most effective detection of fire in high rack storage areas, detectors should be located on the ceiling above each aisle and at intermediate levels in the racks. This is necessary to detect smoke that may be trapped in the racks at an early stage of fire development (when insufficient thermal energy is released to carry the smoke to the ceiling). Earliest detection of smoke can be achieved by locating the intermediate-level detectors adjacent to alternate pallet sections as shown in Figures 13-11 and 13-12.

Detector placement in closed rack storage protects solid storage in which transverse and longitudinal flue spaces are irregular or nonexistent (as for slatted or solid shelved storage). Detectors in open rack storage protect palletized storage or no shelved storage in which regular transverse and longitudinal flue spaces are maintained. Placement of detectors should be according to detector manufacturer's instructions.

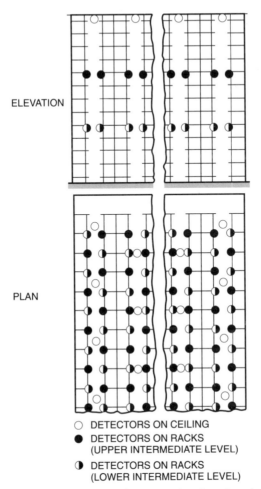

ELEVATION

PLAN

○ DETECTORS ON CEILING
● DETECTORS ON RACKS
  (UPPER INTERMEDIATE LEVEL)
◑ DETECTORS ON RACKS
  (LOWER INTERMEDIATE LEVEL)

Figure 13-11. Detector placement in closed rack storage.

Smoke Detectors for Control of Smoke Spread

Smoke detectors can be used to prevent smoke spread by initiating control of fans, dampers, doors, and other equipment. Detectors for this use can be classified in two ways: (1) area detectors, which are installed in the related smoke compartments; or (2) detectors that are installed in the air duct systems.

Smoke-activated devices are also used to automatically close smoke doors in buildings to limit the spread of smoke in case of fire. Separate corridor ceiling-mounted smoke detectors can be connected to electrically operated hold-open devices on the doors. Smoke detectors can also be built into the door closure units.

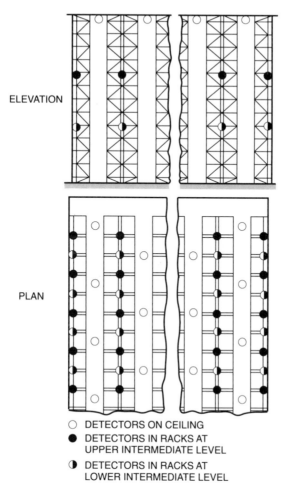

ELEVATION

PLAN

○ DETECTORS ON CEILING
● DETECTORS IN RACKS AT
  UPPER INTERMEDIATE LEVEL
◑ DETECTORS IN RACKS AT
  LOWER INTERMEDIATE LEVEL

Figure 13-12. Detector placement in open rack storage.

Area smoke detectors that are located within a smoke compartment for open area coverage can initiate control of smoke spread by operating doors, dampers, and other equipment.

Smoke Detectors in the Air Duct System

Air duct smoke detectors are installed either in or on return air ducts or at the return air openings of HVAC systems in buildings. This prevents recirculation of smoke from a fire through the HVAC system within the building. Upon detection of a fire, the associated control system initiates an alarm and either (1) shuts down the circulating blowers or (2) switches them to a smoke exhaust mode. Detectors installed in air duct systems should not be considered substitutes for open area protection, because smoke-laden air can be diluted by clean air from other parts of the building, and the smoke may not be drawn into the HVAC system when the system is shut down.

Supply Air System: If detection of smoke in the supply air system is required, the detector(s) listed for the air velocity present can be located in the supply air duct downstream of both the fan and the filters. Smoke detectors should be located within all the smoke compartments served by the supply air system.

Return Air System: If the detection of smoke in the return air system is required it can be accomplished by either: (1) complete area smoke detection (the preferred method) or (2) detector(s) listed for the air velocity present and located at every return air opening within the smoke compartment. In the second case, the detector can also be located where the air leaves each smoke compartment, or in the duct system before the air enters the return air system common to more than one smoke compartment. (See Figures 13-13 and 13-14.)

Additional smoke detectors are not required to be installed in ducts where the air duct system passes through other smoke compartments not served by the duct. (See Figure 13-15.)

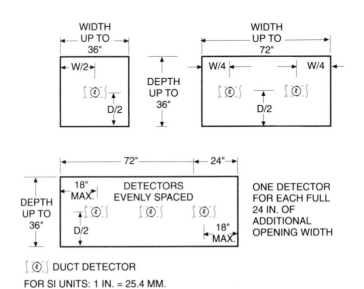

⌠⟨ℓ⟩⌡ DUCT DETECTOR

FOR SI UNITS: 1 IN. = 25.4 MM.

Figure 13-13. Detector spacing in return air system.

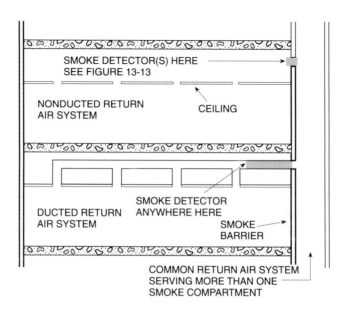

Figure 13-14. Detector location in return air system (selective operation of equipment).

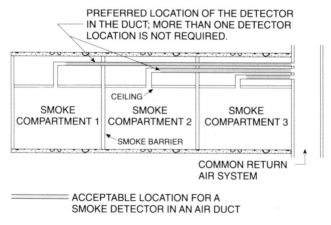

Figure 13-15. Detector location in duct passing through compartments not served by a duct.

Detectors should be readily accessible for cleaning and mounted in accordance with the manufacturer's recommendations. Access doors or panels to the detector should be provided. Detectors mounted outside of a duct employing sampling tubes for transporting smoke from inside the duct to the detector should be arranged to permit verification of air flow from the duct to the detector.

Detectors should be suitable for proper operation over the complete range of air velocities, temperature, and humidity expected at the detector when the air handling system is operating. All penetrations of a return air duct in the vicinity of detectors installed on/in an air duct should be sealed. This prevents entrance of outside air and possible dilution or redirection of smoke within the duct.

Detectors mounted in/on return air ducts should be located at least six duct widths downstream from any duct openings, deflection plates, sharp bends, or branch connection.

If it is physically impossible to locate the detector in this manner, the detector can be positioned closer than the required six duct widths (but as far as possible from the opening, bend, or deflection plate so that smoke can still adequately be detected in the air stream).

The detectors should be installed as shown in Figure 13-13 up to 12 in. (305 mm) in front of or behind the return air opening. Detectors should be spaced according to these dimensions:

1. Width:
   To 36 in. (0.9 m) — One detector centered in opening
   To 72 in. (1.8 m) — Two detectors located at the $\frac{1}{4}$ points of the opening
   Over 72 in. (1.8 m) — One additional detector for each full 24 in. (0.6 m) of opening.
2. Depth: The number and spacing of the detector(s) in the depth (vertical) of the opening should be the same as those given for the width (horizontal) above.
3. Orientation: Detectors should be oriented in the most favorable position for smoke entry, respective to the direction of air flow. The path of a projected beam-type detector across the return air openings should be considered equivalent in coverage to a row of individual detectors.

Smoke Detectors for Door Release Service

When smoke detectors are used to close doors in order to prevent smoke from traveling from one side of the door to the other, they should be mounted and located as shown in Figures 13-16 and 13-17. Additional smoke detectors may be needed for more complex doorways.

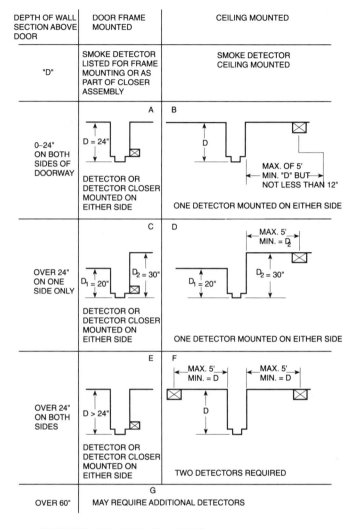

FOR SI UNITS: 1 FT = 0.30 M; 1 IN. = 25.4 MM.

Figure 13-16. Detector mounting for door release service.

ACTUATION OF EXTINGUISHING SYSTEMS

Automatic fire detectors are often used to actuate extinguishing systems, e.g., pre-action and deluge sprinkler systems, combined dry-pipe and pre-action sprinkler systems, and halon systems. Certain conditions should apply when automatic detection systems are used to actuate extinguishing systems. (See NFPA 13, *Standard for the Installation of Sprinkler Systems*, for more information.)

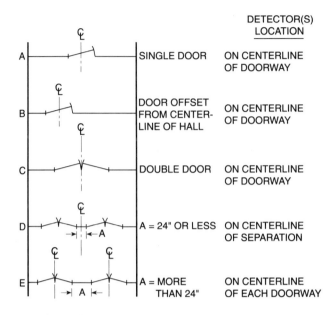

FOR SI UNITS: 1 IN. = 24.5 MM

Figure 13-17. Detector location for door release service.

The following recommendations should be followed in using automatic detection systems to actuate extinguishing systems:

1. The equipment or method used for automatic detection should be listed or approved. Such equipment or methods should detect or indicate heat, smoke, flame, combustible vapors, or any abnormal condition that could identify a fire.

2. Adequate and reliable power supplies should be used. A power supply for an electrical detection system should be independent of the power supply for the protected area. If this separation is not possible, an emergency battery-powered supply with automatic switchover in case of primary power supply failure should be provided.

3. Automatic detection and actuation equipment should be supervised; preferably, indication of equipment failure should be immediately shown at a constantly attended location.

4. System operation should be indicated by audible alarms; the alarms should also alert personnel and indicate failure of a supervised device or piece of equipment. The number of these alarms and their locations should satisfy the local code requirements.

5. Additional alarms should be provided to show that the system has operated, to give personnel ample warning of a discharge from an extinguishing system (especially the halon or high-expansion-foam types), and to indicate failure of supervised devices or equipment.

Pre-Action and Deluge Extinguishing Systems

A pre-action system employs automatic sprinklers attached to a piping system that contains air that may or may not be under pressure, with a supplemental fire detection system installed in the same areas as the sprinklers. Actuation of the fire detection system opens a valve that permits water to flow into the sprinkler piping system and to be discharged from any open sprinklers.

A deluge system employs open sprinklers attached to a piping system and connected to a water supply through a valve that is opened by the operation of a fire detection system installed in the same areas as the sprinklers. When this valve opens, water flows into the piping system and discharges from all sprinklers attached to the piping. Deluge systems may be needed where occupancy conditions or special hazards require quick application of large quantities of water. The fire detection devices or systems for a deluge system with more than 20 sprinklers should be automatically supervised.

Pre-action and deluge systems normally do not have water in the system piping. The water supply is controlled by an automatic valve operated by means of fire detection devices and provided with manual means for operation that are independent of the sprinklers.

Many types of equipment can be included in pre-action and deluge systems, e.g., automatic sprinklers with sprinkler piping and fire detection devices, open sprinklers with fire detection devices, and a combination of open and automatic sprinklers with fire detection devices. The equipment may or may not be supervised.

Fire detection devices should be selected both to ensure operation and to guard against premature operation of sprinklers, based on normal room temperatures and draft conditions. In locations where ambient temperature at the ceiling is high (from heat sources other than fire conditions), heat-responsive devices should be selected that operate at higher than ordinary temperature and that are capable of withstanding the normal high temperature for long periods.

Fire detection devices other than automatic sprinklers should be located and spaced according to their listing by testing laboratories or to manufacturer's specifications. When automatic sprinklers are used as detectors, the distance between detectors and the area per detector should not exceed the maximum permitted for suppression sprinklers as specified by the local code.

Combined Dry-Pipe and Pre-Action Systems

A combined dry-pipe and pre-action sprinkler system employs automatic sprinklers that are attached to a piping system containing air under pressure, and with a supplemental fire detection system installed in the same areas as the sprinklers. Operation of the fire detection system actuates tripping devices that open dry-pipe valves simultaneously without loss of air pressure in the system. Operation of the fire detection system also opens approved air exhaust valves at the end of the feed main, which facilitates the filling of the system with water, preceding the opening of sprinklers. The fire detection system also serves as an automatic fire alarm system.

Combined automatic dry-pipe and pre-action systems should be constructed so that failure of the fire detection system does not prevent the system from functioning as a conventional automatic dry-pipe system. Also, failure of the dry-pipe system of automatic sprinklers should not prevent the fire detection system from properly functioning as an automatic fire alarm system.

Each dry-pipe valve should have an approved tripping device actuated by the fire detection system. Dry-pipe valves should be cross connected through a 1-in. (25.4-mm) pipe connection to permit simultaneous tripping of the dry-pipe valves. The connection should be equipped with a gate valve so that either dry-pipe valve can be shut off or serviced while the other remains in service. (See Figure 13-18.)

Halon Total Flooding Extinguishing Systems

A halon total flooding extinguishing system can be activated by automatic fire detectors. A halon total flooding system is a permanently piped system that uses a limited stored supply of a halon gas under pressure and discharge nozzles to totally flood an enclosed area. The halon releases automatically via a suitable detection system, and extinguishes fires by inhibiting the chemical reaction between fuel and oxygen. (See Figure 13-19.)

This type of extinguishing system is often installed in computer rooms, libraries or archives, or other facility where water would seriously damage contents or equipment. Halon use in sprinklered or unsprinklered computer rooms can protect data in process, reduce

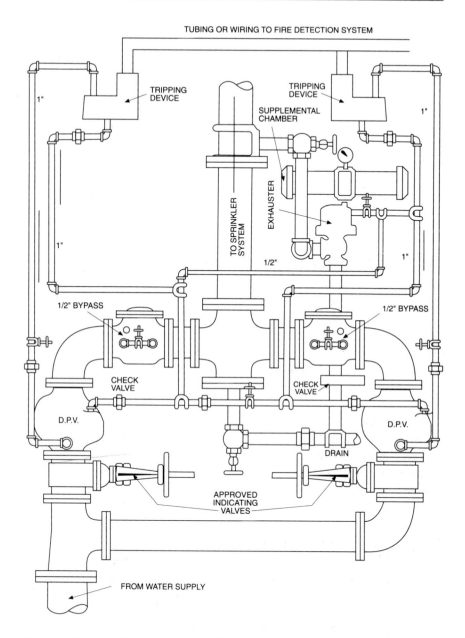

Figure 13-18. Header for combined dry-pipe and pre-action sprinkler system (standard trimmings not shown).

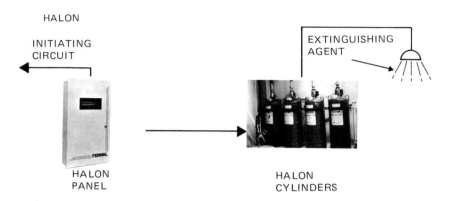

Figure 13-19. Halon discharge in an automatic extinguishing system.

equipment damage, and facilitate return to service. As halon systems are phased out due to damage of the ozone layer, new "clean" agents are being sought as a replacement.

Halon 1301 systems should be automatically actuated by an approved method of detection, meeting the requirements of NFPA 72, *National Fire Alarm Code*. To ensure detection, particular attention should be given to the choice of actuation means, the air flows usually involved in such air-handling systems, and the small heat release under fire conditions.

Where operation of the air conditioning system would exhaust the agent supply, the Halon 1301 system should be interlocked to shut down the air conditioning when the Halon 1301 system is actuated. (Computer equipment does not need to be de-energized prior to discharge.) Alarms should warn area occupants of a discharge or pending discharge.

An automatic detection system should be installed in the space below the raised floor of a computer facility. This additional automatic detection system should meet the above requirements and sound an audible and visual alarm. Air space below the raised floor should be subdivided into areas not exceeding 10,000 sq ft (929 m^2) by tight noncombustible bulkheads.

Cross-Zone Detection for Halon Systems: Although halogenated agent systems generally have operation requirements similar to systems using other types of agents, the use of automatically actuated systems is strongly recommended. This limits the size and severity of fire with which the system must deal, minimizing decomposition of the agent during extinguishment. In a Class A fire, automatic actuation

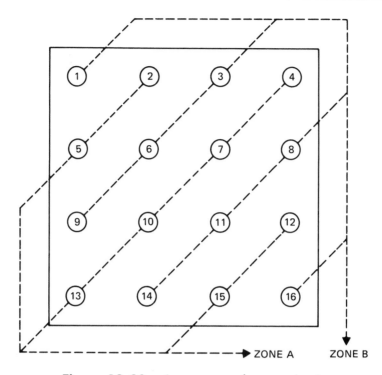

Figure 13-20. A cross-zone detector circuit.

coupled with sensitive detectors can often prevent the fire from becoming deep-seated.

The detector must be sensitive enough to the fuel in question to rapidly respond to the fire at an early stage but not so sensitive as to produce false actuation of the extinguishing system. Sensitivity and reliability in a halon system can be combined by utilizing multiple detectors connected in a double circuit or "cross-zoned" mode of operation. (See Figure 13-20.) Actuation of either zone alone activates local and remote alarms, but does not discharge the extinguishing agent. Actuation of both zones simultaneously causes the agent to discharge. This arrangement prevents a false signal by an individual detector from discharging the entire system. The two zones may contain the same or different types of detectors. Ionization and photoelectric detectors are popularly used in computer areas.

Particular attention must be given to locating detectors within a hazard. The amount and location of combustibles, the listed sensitivity spacing relationship of the detectors considered, and ventilation characteristics of the hazard are all important factors in detector location.

High-Expansion Foam Systems

There are also special considerations in automatically activating high-expansion foam extinguishing systems. This system is a fixed extinguishing system that generates a foam agent for total flooding of confined spaces, and for volumetric displacement of vapor, heat, and smoke. High-expansion foam extinguishes fire by preventing free movement of air, reducing the oxygen concentration at the fire, and cooling.

INSTALLATION AND WIRING FOR FIRE PROTECTIVE SIGNALING CIRCUITS

Installation wiring for a fire protective signaling system should be installed in accordance with the local electrical code. Many times, Article 760 of NFPA 70, *National Electrical Code*, or some variation will be specified.

In wiring a fire alarm system, calculation of proper wire size is critical. The correct wire ensures that the conductor will carry the assigned load and prevents an excessive drop in voltage between the fire alarm control panel and other system components.

There are two separate and distinct circuit classifications: (1) power limited and (2) nonpower limited. (See Article 760 of NFPA 70, *National Electrical Code*.) A power-limited circuit is one in which the power is limited, the circuit supervised, and the circuit durably marked where plainly visible at terminations. If the circuit is not so marked, it must be considered a nonpower-limited circuit. These circuits should never be intermixed, or in the same cable, raceway or same enclosure unless they are separated by a barrier or a mechanically defined separation of at least 2 in. (50 mm). Power-limited and nonpower-limited circuits should also not share the same cable, raceway or enclosure with electric power and lighting circuits, except that power circuits connected to the same equipment can be located in the same enclosure with power-limited conductors and in the same cable and raceway with nonpower-limited circuits.

Power-limited circuits may be intermixed with Class 3 signaling circuits and with Class 2 signaling circuits and communication circuits when their insulation is at least that which is required for the power-limited circuit. Nonpower-limited circuits may be intermixed with Class 1 signaling circuits.

Once a circuit is designated as either power limited or nonpower limited, it must be installed accordingly for its entire length — i.e., a circuit cannot be nonpower limited for a portion of its run and power limited for the remainder.

For example: a smoke detector with both alarm signaling contacts and additional contacts for controlling a local function (such as door release or elevator capture) should be installed so that all connected circuits are either all nonpower limited/Class 1 or all power limited/Class 2/Class 3. This means that the classification of the controlled circuit determines the classification of the fire alarm circuit(s) and the wiring methods and materials that are permitted. Since the fire alarm circuits can terminate at the fire alarm control unit or at terminals adjacent to others for fire alarm circuits that do not have a 2-in. (50-mm) separation or adequate barrier, classifications of most of the other fire alarm circuits in the same control unit enclosure also depend upon the classification of the elevator capture or door release circuit.

Conductors

Requirements below for number of conductors in boxes are taken from Section 370-6 of NFPA 70, *National Electrical Code*. (Table numbers for this text have been substituted for numbers used in NFPA 70.)

There should be boxes of sufficient size to provide free space for all conductors enclosed therein. The maximum number of conductors that should be permitted in standard boxes is shown in Table 13-1.

Table 13-1 must apply where no fittings or devices, such as fixture studs, cable clamps, hickeys, switches, or receptacles, are contained in the box and where no grounding conductors are part of the wiring within the box. Where one or more of these types of fittings, such as fixture studs, cable clamps, or hickeys, are contained in the box, the number of conductors shown in the table must be reduced by one for each type of fitting; an additional deduction of one conductor must be made for each strap containing one or more devices; and a further deduction of one conductor must be made for one or more grounding conductors entering the box. Where a second set of equipment grounding conductors is present in the box, then an additional deduction of one conductor must be made. A conductor running through the box must be counted as one conductor, and each conductor originating outside of the box and terminating inside the box is counted as one conductor. Conductors, no part of which leaves the box, must not be counted. The volume of a wiring enclosure (box) must be the total volume of the assembled sections, and, where used, the space provided by plaster rings, domed covers, extension rings, etc., that are marked with their volume in cubic inches (mm^3), or are made from boxes the dimensions of which are listed in the table.

TABLE 13-1. Standard Metal Boxes

Box Dimension, Inches Trade Size or Type	Min. Cu. In. Cap.	Maximum Number of Conductors						
		No. 18	No. 16	No. 14	No. 12	No. 10	No. 8	No. 6
4 x 1-1/4 Round or Octagonal	12.5	8	7	6	5	5	4	0
4 x 1-1/2 Round or Octagonal	15.5	10	8	7	6	6	5	0
4 x 2-1/8 Round or Octagonal	21.5	14	12	10	9	8	7	0
4 x 1-1/4 Square	18.0	12	10	9	8	7	6	0
4 x 1-1/2 Square	21.0	14	12	10	9	8	7	0
4 x 2-1/8 Square	30.3	20	17	15	13	12	10	6*
4 -11/16 x 1-1/4 Square	25.5	17	14	12	11	10	8	0
4-11/16 x 1-1/2 Square	29.5	19	16	14	13	11	9	0
4-11/16 x 2-1/8 Square	42.0	28	24	21	18	16	14	6
3 x 2 x 1-1/2 Device	7.5	5	4	3	3	3	2	0
3 x 2 x 2 Device	10.0	6	5	5	4	4	3	0
3 x 2 x 2-1/4 Device	10.5	7	6	5	4	4	3	0
3 x 2 x 2-1/2 Device	12.5	8	7	6	5	5	4	0
3 x 2 x 2-3/4 Device	14.0	9	8	7	6	5	4	0
3 x 2 x 3-1/2 Device	18.0	12	10	9	8	7	6	0
4 x 2-1/8 x 1-1/2 Device	10.3	6	5	5	4	4	3	0
4 x 2-1/8 x 1-7/8 Device	13.0	8	7	6	5	5	4	0
4 x 2-1/8 x 2-1/8 Device	14.5	9	8	7	6	5	4	0
3-3/4 x 2 x 2-1/2 Masonry Box/Gang	14.0	9	8	7	6	5	4	0
3-3/4 x 2 x 3-1/2 Masonry Box/Gang	21.0	14	12	10	9	8	7	0
FS—Minimum Internal Depth 1-3/4 Single Cover/Gang	13.5	9	7	6	6	5	4	0
FD—Minimum Internal Depth 2-3/8 Single Cover/Gang	18.0	12	10	9	8	7	6	3
FS—Minimum Internal Depth 1-3/4 Multiple Cover/Gang	18.0	12	10	9	8	7	6	0
FD—Minimum Internal Depth 2-3/8 Multiple Cover/Gang	24.0	16	13	12	10	9	8	4

*Not to be used as a pull box. For termination only.

For combination of conductor sizes shown in Table 13-1, the maximum number of conductors permitted must be computed using the volume per conductor listed in Table 13-2, with the deductions provided for above, and these volume deductions must be based on the largest conductor entering the box. The maximum number and size of conductors listed in Table 13-1 must not be exceeded.

TABLE 13-2. Volume Required Per Conductor

Size of Conductor	Free Space Within Box for Each Conductor
No. 18	1.50 cu in.
No. 16	1.75 cu in.
No. 14	2.00 cu in.
No. 12	2.25 cu in.
No. 10	2.50 cu in.
No. 8	3.00 cu in.
No. 6	5.00 cu in.

Boxes 100 cu. in. (1639 cm^3) or less other than those described in Table 13-1, conduit bodies having provision for more than two conduit entries, and nonmetallic boxes must be durably and legibly marked by the manufacturer with their cubic-inch (cm^3) capacity. The maximum number of conductors permitted must be computed using the volume per conductor listed in Table 13-2, with the deductions provided for above, and these volume deductions must be based on the largest conductor entering the box. Boxes described in Table 13-1 that have a larger cubic-inch (cm^3) capacity than is designated in the table must be permitted to have their cubic-inch (cm^3) capacity marked as required by this section and the maximum number of conductors permitted must be computed using the volume per conductor listed in Table 13-2.

Conduit bodies enclosing No. 6 conductors or smaller must have a cross-sectional area not less than twice the cross-sectional area of the largest conduit to which it is attached. (See Chapter 9, Table 1 of NFPA 70, *National Electrical Code*.)

Conduit bodies having provisions for fewer than three conduit entries must not contain splices, taps, or devices unless they comply with the provisions listed in the preceding paragraph and are supported in a rigid and secure manner.

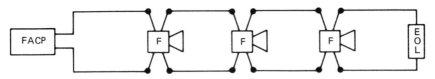

EOL = END-OF-LINE RESISTOR
F = ALARM INDICATING APPLIANCE
FACP = FIRE ALARM CONTROL PANEL

Figure 13-21. Correct wiring method for typical audible signal circuit.

The following information is derived from Article 760 of NFPA 70, *National Electrical Code*, and is meant to be a general guide to circuit installation and wiring. For fire alarm system installation and wiring, a designer should consult and review NFPA 70. Circuits and equipment for fire alarm signaling systems concerning fire spread, ducts, plenums and other air-handling spaces, hazardous (classified) locations, and corrosive, damp or wet locations should comply with NFPA 70.

Fire protective signaling circuits should be identified at terminal and junction locations in a manner that will prevent unintentional interference with the signaling circuit during testing and servicing. Fire protective signaling line circuits that extend aerially beyond one building should meet the requirements of NFPA 70, *National Electrical Code*.

Fire protective signaling circuits and equipment should be grounded except for dc power-limited fire protective signaling circuits having a maximum current of 0.030 amperes. The circuit should also be electrically supervised so that a trouble signal indicates the occurrence of a single open or a single ground fault on any installation wiring circuit that would prevent proper alarm operation.

Interconnecting circuits of household fire warning equipment wholly within a dwelling unit do not need to be supervised.

Wiring for alarm indicating appliances should be similar to that for alarm initiating devices to maintain supervision. Typical wiring circuits are shown in Figures 13-21 and 13-22.

Nonpower-Limited Fire Protective Signaling Circuits

The power supply of nonpower-limited fire protective signaling circuits should comply with the applicable code, and the output voltage should not exceed 600 volts, nominal.

Overcurrent Protection: Conductors should be protected against overcurrent. Overcurrent protection should not exceed 7 amperes for No.

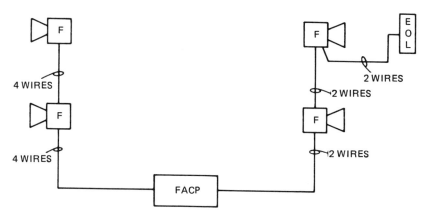

EOL = END-OF-LINE RESISTOR
F = ALARM INDICATING APPLIANCE
FACP = FIRE ALARM CONTROL PANEL

Figure 13-22. Single audible circuit with two risers.

18 conductors, and 10 amperes for No. 16. Other overcurrent protection may be required by NFPA 70, *National Electrical Code*. Overcurrent devices should be located at the point where the conductor to be protected receives its power supply.

Conductors of Different Circuits: Class 1 and nonpower-limited fire protective signaling circuits (ac or dc individual circuits) may occupy the same enclosure, cable, or raceway, provided all conductors are insulated for the maximum voltage of any conductor in the enclosure or raceway. Power supply and fire protective signaling circuit conductors should be in the same enclosure, cable, or raceway only when connected to the same equipment.

Copper Conductors: Copper conductors of Nos. 18 and 16 can be used, provided they supply loads that do not exceed the recommended ampacities and are installed in a raceway or a listed cable. Insulation on conductors should be suitable for 600 volts. Conductors larger than No. 16 should comply with NFPA 70, *National Electrical Code*. No. 16 and No. 18 conductors must be as specified in NFPA 70. Conductors with other type and thickness of insulation can be permitted if listed for nonpower-limited fire protective signaling circuit use.

Conductors should be solid or bunch-tinned (bonded) stranded copper. Two alternative materials are:

1. Stranded copper with a maximum of 7 strands for sizes 16 and 18.
2. Stranded copper with a maximum 19 strands for sizes 14 and larger.

Control of the stranding prevents a single fine strand of wire from completing a circuit, making it appear the circuit is operational. The fine strand could remain intact under the current used for the supervisory signal but burn open when required to carry the higher current of an alarm signal.

Wire types PTF and PAF should be permitted only for high-temperature applications between 194°F (90°C) and 482°F (250°C).

Multiconductor Cable: A multiconductor cable having two or more Nos. 16 or 18 solid or stranded (maximum of 7 strands) copper conductors listed for this use is permitted. This type of cable can be used on fire protective signaling circuits operating at 150 volts or less. The multiconductor cable should be installed in a raceway or exposed in accordance with the requirements of NFPA 70, *National Electrical Code.* Surface-mounted cable should not be permitted within 7 ft (2.13 m) of the floor.

Conductors in Raceways, Cable Trays, and Cables, and Derating: Where only nonpower-limited fire protective signaling circuits and Class 1 circuits are in a raceway, the number of conductors should be determined in accordance with NFPA 70, *National Electrical Code.* Derating factors should apply if such conductors carry continuous load.

Where power supply conductors and fire protective signaling circuit conductors are permitted in a raceway, the number of conductors should be determined in accordance with NFPA 70, *National Electrical Code.* The derating factors specified in NFPA 70 should apply as follows:

1. To all conductors when the fire protective signaling circuit conductors carry continuous loads, and there are more than three conductors.
2. To the power supply conductors only, when the fire protective signaling circuit conductors do not carry continuous loads, and where there are more than three power supply conductors.

Fire protective signaling circuit conductors installed in cable trays must comply with NFPA 70.

Power-Limited Fire Protective Signaling Circuits

As specified in Table 13-3 for ac circuits and Table 13-4 for dc circuits, the power for power-limited fire protective signaling circuits should be either inherently limited (requiring no overcurrent protection) or limited (by a combination of a power source and overcurrent protection).

Supervision: Either a trouble or alarm signal should indicate the occurrence of multiple ground faults or any short circuit fault on the fire alarm system primary (main) power supply, alarm initiating, signaling line, or required alarm indicating circuits that would prevent proper alarm operation.

The circuit should be durably marked where plainly visible at terminations to indicate that it is a power-limited fire protective signaling circuit. Where overcurrent protection is required, the overcurrent protective devices should not be interchangeable with devices of higher ratings. The overcurrent should be permitted as an integral part of the power supply. Overcurrent devices, where required, should be located at the point where the conductor to be protected receives its supply.

Wiring Methods

Conductors and equipment on the supply side of overcurrent protection, transformers, or current-limiting devices should be installed in accordance with appropriate requirements. Transformers or other devices supplied from power supply conductors should be protected by an overcurrent device rated not over 20 amperes.

Input leads of a transformer or other power sources supplying power-limited fire protective signaling circuits can be smaller than No. 14 but not smaller than No. 18, if they are not over 12 in. (305 mm) long and are insulated properly.

Circuits on the load side of overcurrent protection, transformers, and current-limiting devices should have specified wiring types and use those wiring methods and materials. Nonpower-limited circuits should normally comply with Chapter 3 of NFPA 70, *National Electrical Code*, or they can use the multiconductor cables described above. Also, conductors should be solid copper or bunch-tinned (bonded) stranded copper. Power-limited circuits can be reclassified and installed as nonpower-limited circuits under certain conditions.

Power-limited circuit conductors and cables should be installed as follows:

TABLE 13-3. Power Limitations for Alternating-Current Fire Protective
Signaling Circuits

	Inherently Limited Power Source (Overcurrent protection not required)		
Circuit Voltage V_{max} (Note 1)	0 -20	Over 20 -30	Over 30 -100
Power Limitation $(VA)_{max}$ (Note 1) (Volt-Amps)	—	—	—
Current Limitation I_{max} (Note 1) (Amps)	8.0	8.0	$150/V_{max}$
Maximum Overcurrent Protection (Amps)	—	—	—
Power Source Maximum Nameplate Ratings { VA (Volt-Amps) Current (Amps)	$5.0 \times V_{max}$ 5.0	100 $100/V_{max}$	100 $100/V_{max}$

Note 1. V_{max}: Maximum output voltage regardless of load with rated input applied.

I_{max}: Maximum output current after one minute of operation under any noncapacitive load, including short circuit, and with overcurrent protection bypassed if used.

$(VA)_{max}$: Maximum volt-ampere output regardless of load and overcurrent protection bypassed if used.

Note 2.   If the power source is a transformer, $(VA)_{max}$ is 350 or less when V_{max} is 15 or less.

1. In raceways or exposed on surface of ceiling and sidewalls or "fished" in concealed spaces. Where installed exposed, cable should be adequately supported and terminated in approved fittings and installed in such a way that maximum protection against physical damage is afforded by building construction such as baseboards, door frames, ledges, etc. Where located within 7 ft (2.13 m) of the floor, cable should be securely fastened in an approved manner at intervals of not more than 18 in. (457 mm).

2. In metal raceway or rigid nonmetallic conduit when passing through a floor or wall to a height of 7 ft (2.13 m) above the floor unless adequate protection can be afforded by building construction such as detailed in item 1 above, or unless an equivalent solid guard is provided.

3. In rigid metal conduit, intermediate metal conduit, or electrical metallic tubing when installed in hoistways.

TABLE 13-3. Power Limitations for Alternating-Current Fire Protective Signaling Circuits (Continued)

Not Inherently Limited Power Source (Overcurrent protection required)		
0 -20	Over 20 -100	Over 100 -150
250 (see Note 2)	250	NA
$1000/V_{max}$	$1000/V_{max}$	1.0
5.0	$100/V_{max}$	1.0
$5.0 \times V_{max}$ 5.0	100 $100/V_{max}$	100 $100/V_{max}$

Note 1. V_{max}: Maximum output voltage regardless of load with rated input applied.

I_{max}: Maximum output current after one minute of operation under any noncapacitive load, including short circuit, and with overcurrent protection bypassed if used.

$(VA)_{max}$: Maximum volt-ampere output regardless of load and overcurrent protection bypassed if used.

Note 2. If the power source is a transformer, $(VA)_{max}$ is 350 or less when V_{max} is 15 or less.

Single and multiconductor power-limited fire protective signaling circuit cables installed as wiring within buildings should be listed Type FPL listed as being resistant to the spread of fire. In addition, where the cables are in a vertical run in a shaft, and where the cables are installed in ducts, plenums, and other air-handling spaces other requirements in NFPA 70, *National Electrical Code*, may apply.

One method of defining resistance to the spread of fire is that the cables do not spread fire to the top of the tray in the vertical tray flame test in UL 1581, *Standard for Electrical Wires, Cables, and Flexible Cords*, except where the cables are enclosed in a raceway or noncombustible tubing or in nonconcealed spaces where the exposed length of cable does not exceed 10 ft (3.05 m).

Single and multiconductor power-limited fire protective signaling circuit cables in a vertical run in a shaft should be of a fire resistant type

TABLE 13-4. Power Limitations for Direct-Current Fire Protective Signaling Circuits

	Inherently Limited Power Source (Overcurrent protection not required)			
Circuit Voltage V_{max} (Note 1)	0 -20	Over 20 -30	Over 30 -100	Over 100 -250
Power Limitation $(VA)_{max}$ (Note 1) (Volt-Amps)	—	—	—	—
Current Limitation I_{max} (Note 1) (Amps)	8.0	8.0	$150/V_{max}$	0.030
Maximum Overcurrent Protection (Amps)	—	—	—	—
Power Source Maximum Nameplate Ratings $\}$ VA (Volt-Amps) Current (Amps)	$5.0 \times V_{max}$ 5.0	100 $100/V_{max}$	100 $100/V_{max}$	$0.030 \times V_{max}$ 0.030

Note 1. V_{max}: Maximum output voltage regardless of load with rated input applied.

I_{max}: Maximum output current after one minute of operation under any noncapacitive load, including short circuit, and with overcurrent protection bypassed if used.

$(VA)_{max}$: Maximum volt-ampere output regardless of load and overcurrent protection bypassed if used.

Note 2.   If the power source is a transformer, $(VA)_{max}$ is 350 or less when V_{max} is 15 or less.

with characteristics capable of preventing the carrying of fire from floor to floor, except where the cables are encased in noncombustible tubing or are located in a fireproof shaft having firestops at each floor.

Single and multiconductor power-limited fire protective signaling circuit cables and equipment installed in ducts or plenums or other spaces used for environmental air should also be installed in accordance with other sections of NFPA 70, *National Electrical Code*.

Low-smoke-producing cables may be used, and can be defined by establishing an acceptable value of the smoke produced by test to a maximum peak optical density of 0.5 and a maximum average optical density of 0.15. Similarly, fire resistant cables can be defined as having a maximum allowable flame travel distance of 5 ft (1.52 m) when tested to NFPA 262, *Standard Method of Test for Fire and Smoke Characteristics of Wires and Cables*.

Some conductors and cables on the load side of overcurrent protection, transformers, and current-limiting devices should be separated.

TABLE 13-4. Power Limitations for Direct-Current Fire Protective Signaling Circuits (continued)

Not Inherently Limited Power Source (Overcurrent protection required)		
0 -20	Over 20-100	Over 100 - 150
250 (see Note 2)	250	NA
$1000/V_{max}$	$1000/V_{max}$	1.0
5.0	$100/V_{max}$	1.0
$5.0 \times V_{max}$ 5.0	100 $100/V_{max}$	100 $100/V_{max}$

Note 1. V_{max}: Maximum output voltage regardless of load with rated input applied.

I_{max}: Maximum output current after one minute of operation under any noncapacitive load, including short circuit, and with overcurrent protection bypassed if used.

$(VA)_{max}$: Maximum volt-ampere output regardless of load and overcurrent protection bypassed if used.

Note 2.   If the power source is a transformer, $(VA)_{max}$ is 350 or less when V_{max} is 15 or less.

Power-limited circuits should be separated at least 2 in. (50.8 mm) from conductors of any electric light, power, Class 1, or nonpower-limited fire protective signaling circuits. Separation is not required:

1. Where the electric light, power, Class 1, or nonpower-limited fire protective signaling circuit conductors are in raceway or in metal-sheathed, metal-clad, nonmetallic-sheathed, or Type UF cables, or

2. Where the power-limited circuit conductors are permanently separated from the conductors of the other circuits by a continuous and firmly fixed nonconductor (such as porcelain tubes or flexible tubing in addition to the insulation on the wire).

Power-limited circuits should not be placed in any enclosure, raceway, cable, compartment, outlet box, or similar fitting containing conductors of electric light, power, Class 1, or nonpower-limited fire protective signaling circuits except:

1. Where the conductors of the different systems are separated by a partition, or
2. Where conductors in outlet boxes, junction boxes, or similar fittings or compartments where power supply conductors are introduced solely for supplying power to the power-limited fire protective signaling system to which the other conductors in the enclosure are connected.

Power-limited circuits should be separated by not less than 2 in. (50.8 mm) from electric light, power, Class 1, or nonpower-limited fire protective signaling circuit conductors run in the same shaft except:

1. Where the conductors of either the electric light, power, Class 1, the nonpower-limited fire protective signaling circuits, or the power-limited fire protective signaling circuits are encased in noncombustible tubing; or
2. Where the electric light, power, Class 1, or the nonpower-limited fire protective signaling circuit conductors are in a raceway or are in metal-sheathed, metal-clad, nonmetallic-sheathed, or Type UF cables.

The requirements for separation from electric light, power, Class 1, and nonpower-limited fire protective signaling circuits in NFPA 70, *National Electrical Code*, should apply even if the power-limited circuits are wired using nonpower-limited circuit wiring methods.

Cables and conductors of two or more power-limited fire protective signaling circuits or Class 3 circuits may be permitted in the same cable, enclosure, or raceway. Conductors of one or more Class 2 circuits may be within the same cable, enclosure, or raceway with conductors of power-limited fire protective signaling circuits, provided that the insulation of the Class 2 circuit conductors in the cable, enclosure, or raceway is at least that which is required by the power-limited fire protective signaling circuits.

Conductors and Cables

Conductors and cables for use with power-limited fire protective signaling circuits should be listed for this use and meet or exceed the following requirements.

1. Conductors should not be smaller than No. 16 for single conductor, No. 19 for two or three conductor, No. 22 for four or five conductor, and No. 24 for six or more conductor multiconductor cables.
2. Cables should be listed as being suitable for Class 3, power-limited fire protective signaling, or communication circuits.

3. The cable should have a voltage rating of not less than 300 volts and the jacket compound should have a high degree of abrasion resistance.

4. Coaxial cables should have a minimum No. 22 AWG copper or 30 percent minimum conductivity copper-covered steel center conductor, with an overall insulation rated at 300 volts, an overall metallic shield covered by a flame-retardant nonmetallic jacket having a minimum thickness not less than 35 mils nominal (30 mils minimum), and a high degree of abrasion resistance.

5. Listed nonconductive and conductive optical fiber cables can also be used.

6. Listed power-limited fire protective signaling cables should be marked. (See Table 13-5.)

TABLE 13-5. Cable Markings

Cable Marking	Type	NEC Reference
FPL	Power-limited fire alarm cable	760-51(f) and 760-53(c)
FPLP	Power-limited fire alarm plenum cable	760-51(d) and 760-53(a)
FPLR	Power-limited fire alarm riser cable	760-51(d) and 760-53(b)
Cable Marking	Type	NEC Reference
OFC	Conductive optical fiber general purpose cable	770-51(d) and 770-53(c)
OFCP	Conductive optical fiber plenum cable	770-51(a) and 770-53(a)
OFCR	Conductive optical fiber riser cable	770-51(b) and 770-53(b)
OFN	Nonconductive optical fiber general purpose cable	770-51(d) and 770-53(c)
OFNP	Nonconductive optical fiber plenum cable	770-51(a) and 770-53(a)
OFNR	Nonconductive optical fiber riser cable	770-51(b) and 770-53(b)

NOTE: See the referenced NFPA 70, *National Electrical Code* (NEC) sections for permitted uses.

Current-Carrying Continuous Line-Type Fire Detectors

Listed continuous line-type fire detectors, including insulated copper tubing of pneumatically operated detectors that are employed for both detection and carrying signaling currents, can be used in circuits having power-limiting characteristics.

Optical Fiber Cables

Optical fiber cables along with electrical conductors can also be used for fire alarm signal transmission. Optical fiber cables transmit light for control, signaling, and communications through an optical fiber and can be grouped into three types: (1) nonconductive cables contain no metallic members and no other electrically conductive materials; (2) conductive cables contain noncurrent-carrying conductive members, such as metallic strength members and metallic vapor barriers; and (3) hybrid cables contain optical fibers and current-carrying electrical conductors and should be classified as electrical cables in accordance with the type of electrical conductors.

Optical Fibers and Electrical Conductors: Optical fibers can be contained within the same hybrid cable for electric light, power, or Class 1 circuits operating at 600 volts or less only where the functions of the optical fibers and the electrical conductors are associated. Nonconductive optical fiber cables can occupy the same raceway or cable tray with conductors for electric light, power, or Class 1 circuits operating at 600 volts or less. Conductive and hybrid optical fiber cables should not occupy the same raceway or cable tray with conductors for electric light, power, or Class 1 circuits.

Nonconductive optical fiber cables should not be permitted to occupy the same cabinet, panel, outlet box, or similar enclosure housing the electrical terminations of an electric light, power, or Class 1 circuit unless nonconductive optical fiber cable is functionally associated with the electric light, power, or Class 1 circuit.

Occupancy of the same cabinet, panel, outlet box, or similar enclosure is allowed where nonconductive optical fiber cables are installed in factory- or field-assembled control centers. Nonconductive optical fiber cables should be permitted with circuits exceeding 600 volts only in industrial establishments where conditions of maintenance and supervision ensure that only qualified persons will service the installation.

Optical fibers can be in the same cable; and conductive and nonconductive optical fiber cables are permitted in the same raceway, cable tray, or enclosure with conductors of any of the following (in accordance with NFPA 70, *National Electrical Code*):

1. Class 2 and Class 3 remote-control, signaling, and power-limited circuits.
2. Power-limited fire protective signaling systems.
3. Communications circuits.
4. Community antenna television and radio distribution systems.

Noncurrent-carrying conductive members of optical fiber cables should be grounded.

Fire Resistance of Optical Fiber Cables: Optical fiber cables installed as wiring within buildings should be listed as being resistant to the spread of fire. Cables in a vertical run in a shaft or installed in ducts, plenums, and other air-handling spaces have other fire resistance requirements as outlined in NFPA 70, *National Electrical Code*. Optical fiber cables may be excepted from requirements where the optical fiber cables are enclosed in raceway or noncombustible tubing, or in nonconcealed spaces where the exposed length of cable does not exceed 10 ft (3.05 m).

Optical fiber cables in a vertical run in a shaft should have either fire resistant characteristics capable of preventing the carrying of fire from floor to floor, or the optical fiber cables should be encased in noncombustible tubing or be located in a fireproof shaft having firestops at each floor.

Optical fiber cables and equipment installed in ducts or plenums or other spaces used for environmental air should also be installed in accordance with the wiring requirements of NFPA 70, *National Electrical Code*. Certain types of listed optical fiber cables with adequate fire resistance and that are low-smoke producing may be allowed for ducts, plenums, and other space used for environmental air.

Low-smoke-producing cables can be defined by establishing an acceptable value of the smoke produced to a maximum peak optical density of 0.5 and a maximum average optical density of 0.15. Similarly, fire resistant cables can be defined as having a maximum allowable flame travel distance of 5 ft (1.52 m) when tested to NFPA 262, *Standard Method of Test for Fire and Smoke Characteristics of Wires and Cables*.

Where exposed to contact with electric light or power conductors, the noncurrent-carrying metallic members of optical fiber cables entering buildings should be grounded as close to the point of entrance as practicable or interrupted as close to the point of entrance as practicable by an insulating joint or equivalent device. The point of entrance should be considered to be at the point of emergence through an exterior wall, a concrete floor slab, or from a rigid metal conduit or an intermediate metal conduit that is grounded.

Listed optical fiber cables should be marked in accordance with Table 13-5.

TRANSIENT AND LIGHTNING PROTECTION

When fire alarm system equipment is submitted to a testing laboratory for listing, each type of equipment undergoes extensive transient tests. Despite this, there may be excessive electrical transients in the building that houses the fire alarm system, and installation of transient protection in the field should be considered.

Most commonly experienced high-voltage transients and interferences are of short duration and/or have low energy content. High-energy transients, although rare and usually caused by direct lightning strikes, tend to be catastrophic to electronic equipment and require elaborate protective measures.

Protective signaling systems are susceptible to many types of electrical interference, including:

1. Lightning strikes, direct or induced surges;
2. Uneven powerline conditions and transients;
3. Transients generated by switching various system components, such as relays, bells, etc.;
4. Interference induced by capacitive, inductive, or electromagnetic coupling to system wiring from nearby motors, neon signs, radio-frequency transmitters, etc.; and
5. Direct coupled transients on system wiring (other than those on powerlines) that are caused by direct or secondary lightning strikes, etc. (usually catastrophic to equipment).

Transient protection can be provided by transient suppressors or other components. To be effective, transient suppressors must conduct more current than the corresponding voltage rise indicates as, for example, a resistor does not. The current increase in a resistor is directly proportional to an increase in the supply voltage to it. Nonlinear devices, such as thyristors and metal oxide varistors (MOV), have current rises to some exponential power of the applied voltage (such as $I = KV^x$ where x, a characteristic of the device, runs typically from 5 to 35). Extreme nonlinear devices, such as zener diodes and spark gaps,

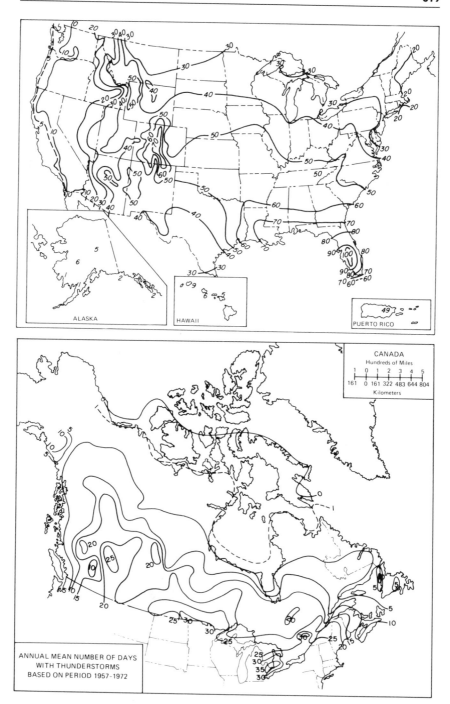

Figure 13-23. Mean annual days with thunderstorms, U.S. and Canada.

are essentially breakover devices in that, when a certain voltage is reached, the current rises dramatically. Various suppressors can be selected for applications, depending upon cost, current-handling capability, response time, etc.

Suppressor selection depends upon detailed device characteristics as they apply to the intended use. Spark gaps are probably the most difficult of suppressors to apply correctly, because their discharge can continue to conduct until the applied voltage is reduced below 25 volts (for commonly available types). In general, protecting each conductor will provide a more reliable system, particularly in lightning-prone areas or where commercial power can destroy electronic apparatus. (See Figure 13-23.) Any circuit that is not sheathed in grounded conduit is highly subject to becoming an antenna, particularly if the circuits are ungrounded. Conservative design requires transient protection on each line to ground.

BIBLIOGRAPHY

NFPA Codes, Standards, Recommended Practices, and Manuals. (See the latest *NFPA Catalog* for availability of current editions of the following documents.)

NFPA 11A, *Standard for Medium- and High-Expansion Foam Systems.*

NFPA 12A, *Standard on Halon 1301 Fire Extinguishing Systems.*

NFPA 13, *Standard for the Installation of Sprinkler Systems.*

NFPA 70, *National Electrical Code.*

NFPA 72, *National Fire Alarm Code.*

NFPA 75, *Standard for the Protection of Electronic Computer/Data Processing Equipment.*

NFPA 90A, *Standard for the Installation of Air Conditioning and Ventilating Systems.*

NFPA 101, *Life Safety Code.*

NFPA 231C, *Standard for Rack Storage of Materials.*

NFPA 262, *Standard Method of Test for Fire and Smoke Characteristics of Wires and Cables.*

NFPA 780, *Lightning Protection Code.*

Additional Reading

UL 1581, *Standard for Electrical Wires, Cables, and Flexible Cords,* Underwriters Laboratories Inc., Northbrook, IL, 1991.

14

Testing

INTRODUCTION

After proper installation, an ongoing test program is the factor that most contributes to the continued efficient operation of a fire alarm signaling system and its components. This chapter describes the importance of testing and the differences between acceptance testing and periodic testing. Test schedules for various types of systems and for automatic fire detectors are also provided. For additional information, see Chapter 6, "Electrical Supervision"; Chapter 8, "Power Supplies;" and Chapter 13, "Installation."

THE IMPORTANCE OF TESTING

All electronic components fail. Basic electronic components of a fire alarm system, then, will also fail at some time. For example: a smoke detector, one of the most important parts of a fire alarm signaling system, can be out of service for long periods unless a functional test is performed.

Failure rates for smoke detectors can be calculated, since the failure rate of standard electronic components is known. "Failure" is expressed in downtime, or lost time in operation, and is measured in hours. The failure rate is further defined as failure other than that caused by loss of power, physical damage, dirt or dust, electrical transients, or actual removal of the system or components. Across the industry, a rate of 4 failures per 1 million hours of operation is considered the maximum design failure rate that can be permitted. This text more conservatively recommends a 3.5 failure rate per 1 million hours. Table 14-1 shows

TABLE 14-1. Probability of Failure and Average Unprotected Time for Typical Ranges of Detector Failure Rates and Test Intervals for Various Service Life Periods

Service Period	1 Year				10 Years			
Failures per million hours	2.0	3.0	3.5	4.0	2.0	3.0	3.5	4.0
Unprotected time (weeks) Test interval								
1 week	2.5	2.5	2.5	2.5	2.7	2.8	2.9	3.0
2 weeks	3.0	3.0	3.0	3.1	3.3	3.4	3.5	3.6
1 month	4.2	4.2	4.2	4.2	4.5	4.7	4.8	5.0
3 months	8.6	8.6	8.7	8.7	9.3	9.7	9.9	10.1
6 months	15.2	15.2	15.3	15.3	16.4	17.1	17.5	17.9
1 year	28.2	28.6	28.6	28.7	30.6	32.0	32.7	33.5
No test	28.2	28.6	28.6	28.7	268	271	273	276
Probability of failure during service period (%)	1.7	2.6	3.0	3.4	16.1	23.1	26.4	29.6

that at the 3.5 rate, 3 percent of the detectors will fail in one year. This failure rate progresses mathematically until at the end of 30 years, 60 percent of the detectors will fail.

Failure rates are important especially when applied to millions of smoke detectors in use in the U.S. A failure rate of 3.5 per million hours of operation for every 10 million smoke detectors translates into 35 detectors that are failing every hour of every day; for every 100 million detectors, 350 are failing every hour of every day. Smoke detectors must be tested regularly to identify those detectors that have not failed.

Without testing, in a one-year period, the detector could be out of service for as long as 28 weeks; 273 weeks for 10 years; approaching 900 weeks (about 17 years) in 30 years. (See Figure 14-1.) By testing the detector only twice a year, i.e., every six months, service downtime is cut in one-half. (See Figure 14-2.) The out-of-service time with a six-month test is, over a 30-year period, only $\frac{1}{40}$th of the time with no test. (See Table 14-1.)

TABLE 14-1. Probability of Failure and Average Unprotected Time for Typical Ranges of Detector Failure Rates and Test Intervals for Various Service Life Periods (continued)

	20 Years				30 Years		
2.0	3.0	3.5	4.0	2.0	3.0	3.5	4.0
3.0	3.2	3.3	3.5	3.2	3.6	3.8	4.0
3.6	3.9	4.0	4.2	3.9	4.3	4.6	4.9
4.9	5.4	5.6	5.8	5.4	6.0	6.4	6.7
10.1	10.9	11.4	11.9	10.9	12.3	13.0	13.8
17.8	19.3	20.1	21.0	19.3	21.7	23.0	24.4
33.3	36.1	37.6	39.3	36.1	40.6	43.0	45.6
550	565	573	581	848	881	898	914
29.6	40.1	45.8	50.4	40.1	54.4	60.1	65.0

With a one-year test interval, if the detector design failure rate is changed from 4 to 2 failures per 1 million hours, the unprotected time would drop from 45 to 36 weeks. However, testing a smoke detector with a failure rate of 4 failures per 1 million hours once a month instead of once a year would reduce the unprotected time from 45 to 6.7 weeks. Figure 14-3 clearly shows that regular testing of smoke detectors is the best way to identify a failed unit, and, by promptly replacing it, reduce "out-of-service time."

All failure rates previously discussed apply equally to the single-station residential smoke detectors and to the system smoke detectors found in a protected building.

SYSTEM MALFUNCTIONS

Testing and maintaining fire alarm signaling systems retains the designed operating characteristics of the system and affirms system operation. Common system deficiencies can be corrected via proper

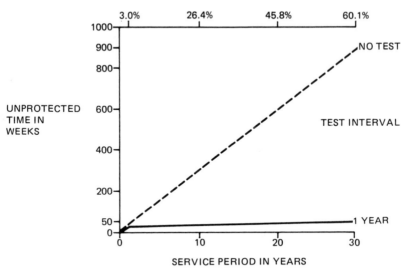

Figure 14-1. Probability of smoke detector failure with no test, and one-year test intervals.

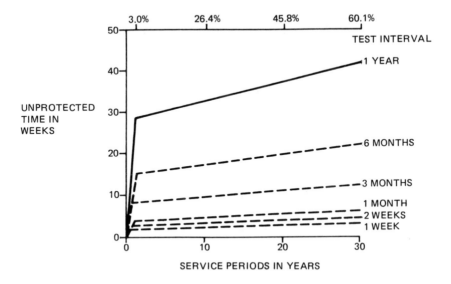

Figure 14-2. Probability of smoke detector failure with different test intervals during service period. (Plot of Table 14-1 with 3.5 failures per million hours.)

TEST FEATURES

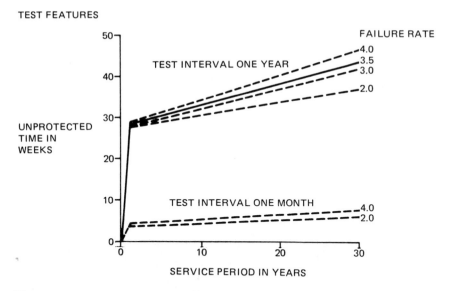

Figure 14-3. Failure rate for smoke detectors with different failure rates—monthly, and annual tests.

testing and maintenance. A study of fire alarm system effectiveness conducted by Los Angeles County (CA) fire chiefs[1] found that, of the more than 46,000 systems surveyed, many connected to central stations did not function properly.

Slightly more than 6,000 systems of those surveyed had failed or malfunctioned during the 12 months preceding the survey. The top ten reasons for failure, ranked in order of frequency, were:

1. Faulty flow switch (30 systems)
2. Lack of maintenance (29)
3. Water in conduit (23)
4. Power failure (18)
5. Telephone line trouble (17)
6. Detector failure (16)
7. Vandalism (not repaired)(11)
8. Battery failure (11)
9. Detectors too sensitive (10)
10. Poor installation (9).

(Other failures were caused by central station problems, lack of detector maintenance, short circuit, equipment failure or incompatibility, technician related, etc.)

Another study[2] concerned fire alarm signaling systems in 45 hotels; 20 hotels had 26 control panels under 3 years of age, and the remaining had 27 new control panels. The study included a visual inspection, functional test of control panels, and electrical supervision test (wire disconnected).

Most control panels that did not operate properly had operational problems, such as burned-out lamps, dead batteries, or inoperative city connections. Many of the system smoke detectors in these hotels that would not alarm were not supervised or had other problems; some fire alarm indicating appliances for the hotel systems were also not supervised while others had burned-out lamps. Table 14-2 provides more details of the test failures in surveyed system components.

The hotel survey revealed further problems that could have been detected through a regular test program. Additional system problems included unsealed riser piping between guest rooms, gaps in smoke doors, and improperly wired sprinkler flow switches (and unsupervised flow switches). Control panels had no trouble signal on transfer to battery, or an inoperative resound circuit; system detector wires were looped around the terminal screw; single-station detectors were missing in some cases or had low-sounding horns; and some audible appliances were missing or improperly wired.

Frequent tests for every major part of the fire alarm system will address failure rates and malfunctions in equipment or system components.

TESTING PROCEDURES

Testing procedures can be classified into two categories: (1) acceptance and re-acceptance testing procedures or (2) periodic equipment and circuit testing procedures.

Acceptance Testing

Every fire alarm signaling system should be completely tested after installation or alteration to ensure that the system meets its design objectives and is fully operational. Qualified personnel, including a fire protection engineer, factory representative of the equipment manufacturer, and an engineer of the installation company should participate in the startup test. A detailed testing procedure should be prepared prior to the test, with all parties becoming familiar with the procedure before the test is actually conducted.

TABLE 14-2. Component Problems in Hotel Systems Survey (% of total components)

Tested Component	System Less than 3 Years	New System
Control Panels		
Alarm condition	19%	7%
Trouble signal	30%	14%
Operational problems	50%	39%
Total Panels	26	27
% with problems	99%	60%
System Smoke Detector		
No alarm	2%	0.9%
Not supervised	24%	0.5%
Mounted in wrong location	1%	0.07%
Other problems	3%	3.3%
Total Detectors	942	1,347
% with problems	30%	3.51%
Alarm Indicating Appliances		
Did not sound properly	6%	0.64%
Not supervised	11%	1.4%
Burned-out lamps	8%	3.5%
Total Appliances	524	1,720
% with problems	25%	5.54%

After installation or alterations, satisfactory tests of the entire system should be made in the presence of a representative of the local authority. All functions of the system should be tested, including operation of the system in various alarm and trouble modes for which it is designed (e.g., open circuit, grounded circuit, power outage, etc.).

As-built drawings and the manufacturer's manual should be available for verifying the agreement between the connected equipment and the as-built drawings. Prior to connection of the equipment, all conductors should be tested as follows. (NOTE: Equipment should not be wired during conductor testing, or equipment damage may result.)

1. Absence of unwanted voltages between circuit conductors and ground that would constitute a hazard or prevent proper system operation should be verified.
2. All conductors other than those intentionally and permanently grounded should be tested for isolation from ground using an insulation testing device such as a "megger." All conductors other than those intentionally connected together should be tested for conductor-to-conductor isolation using an insulation testing device.
3. With each circuit pair short-circuited at the far end of the circuit, circuit resistance should be measured with an ohmmeter and recorded for each circuit shown on the as-built drawings.

The entire system should be tested as follows:

1. The control unit should be tested to verify it is in the normal supervisory condition as detailed in the manufacturer's manual.
2. Each initiating and indicating circuit should be tested to confirm that the integrity of installation conductors is being properly monitored by a suitable response at the control unit. One connection each should be opened at no less than 10 percent of all initiating devices and indicating appliances.
3. Each initiating device and indicating appliance of the system should be tested for alarm operation and proper response at the control unit. All intended functions should be tested in accordance with the manufacturer's manual, including all supplementary functions. Primary (main) power supplies and secondary (standby) power supplies should be tested.

A re-acceptance test should be performed on all equipment and circuits affected by system additions, deletions, or damages after any addition, deletion, or mechanical or electrical damage to the system has occurred, or when conductors are added.

As part of the acceptance check, the local authority should be furnished a certificate of compliance. (See Figure 14-4.) All parts of the certificate should be completed after the system is installed and the installation wiring has been checked, except for Part 2 which should be completed after the operational acceptance tests have been completed. A preliminary copy of the certificate is given to the system owner and, when requested, to the authority(ies) having jurisdiction after completion of the installation wiring tests. A final copy of the certificate is distributed after completion of the operational acceptance tests.

	Protected Property	System Installer	System Supplier	Service Organization
Name				
Address				
Representative				
Telephone				

Location of As-Built Drawings: _____

Location of Owners' Manuals: _____

Location of Test Reports: _____

1. **Certification of System Installation:** (Fill out after installation is complete and wiring checked for opens, shorts, ground faults, and improper branching, but prior to conducting operational acceptance tests.)
   This system installation was inspected by _____
   on _____ and found to comply with the installation requirements of:
   ____ NFPA 72, *National Fire Alarm Code*
   ____ Article 760 of NFPA 70, *National Electrical Code*
   ____ Manufacturer's Instructions
   ____ Other (specify) _____

   Signed _____ Date _____

   Organization _____

2. **Certification of System Operation:**
   All operational features and functions of this system were tested by
   _____ on _____ and found to be operating properly in accordance with the requirements of:
   ____ NFPA 72, *National Fire Alarm Code*
   ____ Job Specifications
   ____ Manufacturer's Instructions
   ____ Other (specify) _____

   Signed _____ Date _____

   Organization _____

   Test Witness for the Authority Having Jurisdiction _____

Figure 14-4. Fire alarm system certificate and description.

3. **Type(s) of System or Service:**

___ Local. If alarm is transmitted to location(s) off premise, list where received: _____

___ Auxiliary. Indicate type of connection: _____
Local Energy ___ , Shunt ___ , Parallel Telephone _____
Location and telephone number for receipt of signals: _____

___ Remote Station.
Location and telephone number for receipt of signals: _____
Alarm: _____
Supervisory: _____

___ Proprietary. If alarms are retransmitted to Public Fire Service Communications Center or Central Station, indicate location and telephone number of the organization receiving alarm. _____

_____

Indicate how alarm is retransmitted _____

___ Emergency Voice/Alarm Service.
Quantity of voice/alarm channels: _____ Single _____ , Multiple (specify) _____
Quantity of speakers installed: _____ Quantity of speaker zones: _____
Quantity of telephones or telephone jacks included in system: _____

4. **Alarm Initiating Devices and Circuits:**

Quantity and style of initiating device circuits connected to system:
Quantity _____ , Style _____ .

Types and quantities of alarm initiating devices installed:
_____ Manual stations: _____ Noncoded, _____ Coded Quantity: _____
_____ Smoke detectors: _____ Ion,   _____ Photo. Quantity: _____
_____ Duct detectors: _____ Ion,   _____ Photo. Quantity: _____
_____ Sprinkler waterflow switches   Quantity: _____
_____ Other: (list) _____   Quantity: _____

Figure 14-4.   Continued.

5. **Alarm Indicating Appliances and Circuits:**
   Quantity of indicating appliance circuits connected to system: _____
   Types and quantities of alarm indicating appliances installed:
   _____ Bells: Size: _____ .   Quantity: _____
   _____ Horns: Quantity: _____
   _____ Chimes: Quantity: _____
   _____ Other: (specify) _____ .   Quantity: _____
   _____ Visible Signals: Type: _____ .   Quantity: _____
          Indicate whether _____ combined with audible,
              or _____ mounted separately.

6. **Supervisory Signal Initiating Devices and Circuits:**
   Quantity and style of supervisory circuits:
      Quantity _____ , Style _____
   Types and quantities of supervisory signal initiating devices installed:
   _____ Sprinkler control valve   Quantity: _____
   _____ Building temperature   Quantity: _____
   _____ Site water temperature   Quantity: _____
   _____ Site water supply level   Quantity: _____

   Electric fire pump:
   _____ Fire pump power   Quantity: _____
   _____ Fire pump running   Quantity: _____

   Engine-driven fire pump:
   _____ Selector in auto. position   Quantity: _____
   _____ Engine or control panel trouble   Quantity: _____
   _____ Fire pump running   Quantity: _____

   Engine-driven generator:
   _____ Selector in auto. position   Quantity: _____
   _____ Control panel trouble   Quantity: _____
   _____ Transfer switch   Quantity: _____
   _____ Engine running   Quantity: _____

   Other supervisory function (specify): _____   Quantity: _____

   _____

   _____

7. **Signaling Line Circuits:**
   Quantity and style of signaling line circuits connected to system:
      Quantity _____ , Style(s) _____

Figure 14-4. Continued.

8. **System Power Supplies**
   a. Primary (main): Nominal voltage: _____ , Amps _____
      Overcurrent protection: Type _____ , Amps _____
      Location _____
   b. Secondary (standby):
      _____ Storage battery: Amp-hr. rating _____
      Calculated capacity to operate system, in hours:
      _____ 24, _____ 60
      _____ Engine-driven generator dedicated to fire alarm
         system:
         Location of fuel storage: _____
   c. Emergency or standby system used as backup to primary power
      supply, instead of using a secondary power supply:
      _____ Emergency system
      _____ Legally required standby system
      _____ Optional standby system

Figure 14-4.   Continued.

Chapter 7 of NFPA 72, *National Fire Alarm Code,* provides information
for performing the installation wiring and operational acceptance tests
required when completing the certificate of compliance. (See Table 14-3.)

Central Station Acceptance Testing: Many agencies can be involved in
the approval of the central station signaling system installation, because
the protection is provided for multiple properties under different own-
erships. These agencies could include the local fire department or fire
prevention bureau, insurers, state regulatory agencies, etc.

All tests and maintenance for any fire alarm signaling system, including a central station system, should be supervised and performed by
qualified personnel. Complete test and maintenance records should be
kept for all systems and components.

Complete information regarding the central station system, including
specifications, wiring diagrams, and floor plans should be submitted for
approval to the local authority prior to installation of equipment or
wiring. All devices, combinations of devices, and equipment constructed
and installed for a central station system should be approved for their
intended purposes and a certificate prepared. (See Figure 14-5.) Upon
completion of a system, a satisfactory test should be made in the presence of a representative of the local authority, if so required.

Property name _____   Central station company_____

Address _____   Station location _____

To: _____   Address _____
   (authority having jurisdiction)

---

1. A contract (dated_____) for test and inspection in accordance with
   NFPA standard(s) no.(s) _____ [edition date(s)] is in effect.

---

2. The system(s) has been installed in accordance with NFPA standard(s)
   no.(s) _____ [edition date(s) _____], includes the devices listed below, and
   has been in service since (date _____).

---

3. Location   Building(s) — name, number, etc.

---

4. Manual fire   (a) Number of coded stations _____
   alarm service   (b) Number of non-coded stations _____ activating
                           fire alarm transmitters
                       (c) Number of combination manual fire alarm and
                           guard's tour coded stations _____
                       (d) Local annunciator — yes   no

---

5. Guard's tour   (a) Number of coded stations _____
   supervisory   (b) Number of non-coded stations _____ activating
   service   transmitters
                       (c) Compulsory guard's tour system consisting
                           of _____ transmitter stations and _____ interme-
                           diate stations.
                       Note: Record combination devices under 4(c) & 5(a)

---

6. Automatic fire   (a) Coverage   full   partial
   detection and   (If partial, indicate locations)
   alarm service   (b) Types of detectors and number of each (for line
                           type, indicate number of circuits)
                       (c) Local annunciator — yes   no
                       (d) Local alarm   — yes   no
                       (e) Number of coded fire signals _____
                           Number of coded trouble signals _____

Figure 14-5. Certification of completion for a central station signaling
system.

7. Sprinkler sys-   (a) Number of coded waterflow signaling
   tem water-   attachments _____
   flow alarm &   Number of waterflow switches _____ activating
   supervisory   transmitters
   service   (b) Number of coded valve supervisory signaling
                        attachments _____
                        Number of valve switches _____
                        activating _____ transmitters
                    (c) Other supervisory service provided
                        Pressure: water   air
                        Temperature: water   room
                        Water level   fire pump: running   power
                    (d) Other fire service provided _____

---

8.   (a) Means of transmission of signals from the pro-
                        tected premises to the central station: McCulloh
                        multiplex   digital alarm communicator
                        two-way rf
                    (b) Means of retransmission of alarms to the public
                        fire service communications center: 1. _____
                        2. _____

---

9. Comments

(Signed) for central station company   Title   Date

Frequency of routine tests and inspections, if other than in accordance with
the referenced NFPA standard(s): _____

System(s) deviations from the referenced NFPA standard(s) are:

_____

_____

_____

(Signed) for central station company   Title   Date

Upon completion of the system(s), satisfactory test(s) witnessed (if required by the
authority having jurisdiction)

By _____
    (representing the authority having jurisdiction)   Title   Date

Figure 14-5.   Continued.

Other Systems: Additional test and maintenance are required for emergency voice/alarm communication systems. After completion of acceptance tests a set of reproducible "as-built" installation drawings, operation and maintenance manuals, test and maintenance schedules, and records for this system should be provided for the owner or a designated representative and kept current.

Audible appliances for the public operating mode of an emergency voice/alarm system should be tested and maintained in accordance with the manufacturer's instructions. Tests should be conducted using a portable sound level meter meeting the Type 2 requirements of ANSI S.1.4, *Specification for Sound Level Meters* ("A"-weighted scale).[3] Audible appliances in the private operating mode should be tested and maintained in accordance with the manufacturer's instructions.

Visible appliances for the public operating mode should be tested and maintained in accordance with the ratings, procedures, and component parts replacement schedule recommended by the manufacturer for their intended use. Visible appliances for the private operating mode should be tested and maintained as rated and specified by the manufacturer or local authority having jurisdiction, including the recommended change schedule for replacement components.

Acceptance Testing for Automatic Fire Detectors: Responsibility for inspections, tests, and maintenance programs should be assigned by the owner to a designated person. People at all locations where the alarm signals or reports should be notified before testing to prevent unnecessary response, and notified when testing has concluded. Any method or device used for testing in an atmosphere or process classified as hazardous should be suitable for such use. Records of all inspections, tests, and maintenance should be kept on the premises for a minimum of five years.

After installation of the system all detectors should be visually inspected to be sure that they are properly located, and connected in accordance with the manufacturer's recommendations.

Heat Detectors: A restorable heat detector and the restorable element of a combination detector should be tested by exposing the detector to a heat source (such as a hairdryer or a shielded heat lamp) until it responds. The detector should reset after each heat test. Precautions should be taken to avoid damage to the nonrestorable fixed temperature element of a combination rate-of-rise/fixed temperature detector.

A pneumatic tube line-type detector should be tested with either a heat source (if a test chamber is in the circuit) or pneumatically with a pressure pump according to manufacturer's instructions. Line- or spot-type nonrestorable fixed temperature heat detectors should be tested mechanically or electrically (not with heat) to verify alarm function. If required for proper performance, the loop resistance of line-type detectors should be measured to determine if it is within acceptable limits for the equipment being used. The resistance should be recorded for future reference.

Detectors with a replaceable fusible alloy element should be tested by first removing the fusible element to determine that the detector contacts operate properly, and then reinstalling the fusible element.

Smoke, Flame, and Other Fire Detectors: An alarm should be initiated at each smoke detector with smoke or other aerosol to demonstrate that smoke can enter the chamber and initiate an alarm.

Most residential smoke detectors have an integral test means permitting an individual to check the system and its sensitivity. Underwriters Laboratories Inc. (UL) Standards 217, *Standard for Safety Single and Multiple Station Smoke Detectors*, and 268, *Standard for Safety Smoke Detectors for Fire Protective Signaling Systems*, require that each smoke detector be provided with either a means of measuring detector sensitivity or a built-in sensitivity test feature. These standards allow the test feature to be either electrical or mechanical, but it is required to simulate a smoke level of no more than 6 percent per ft (0.3 m). Though this exceeds the normal factory-set level of between 0.5 and 3 percent per ft (0.3 m), it is felt that 6 percent per ft is a tolerable level. A more accurate measurement of smoke detectors in the field can be performed using an aerosol generator developed by the National Bureau of Standards (NBS).

Testing a detector, however, does not increase reliability. Tests will only identify detectors that have either failed or have shifted their sensitivity (set point).

Blowing smoke into a detector or using an aerosol can spray to test the detector is not recommended since the density of either cannot be accurately determined. Using these methods could impose a 50 to 60 percent per ft (0.3 m) particle density on a detector and lead one to believe that the detector was still within calibrated limits, when in fact the detector set point could have shifted to an unacceptable level of 30 to 40 percent per ft (0.3 m).

One method used to simulate 6 percent smoke in a photoelectric detector is a calibrated test wire that is built into the detector. This test wire passes through the light path when the test button is pressed. The test wire scatters the same amount of light onto the photosensitive device as when smoke of about 6 percent per ft (0.3 m) density enters the smoke-sensing chamber.

In ionization detectors, various methods are used to simulate smoke. In some detectors, the electrical conductivity of the smoke-sensing chamber is changed by placing a shunt across the chamber. In other detectors, the geometry of the chamber is physically changed, or a voltage is impressed on the circuit in the smoke chamber that would simulate the output voltage with a given density of smoke particles.

The factory calibration method or a calibrated particle generator only should be used to test the sensitivity of smoke detectors, because a specific level of smoke is needed for a proper test.

To test a smoke detector sensitivity range, the detector should be tested using a calibrated test method, the manufacturer's calibrated sensitivity test instrument, listed control equipment arranged for the purpose, or other calibrated sensitivity test method acceptable to the local authority.

Detectors with a sensitivity outside the approved range should be replaced. Detectors listed as field adjustable may be either adjusted within an approved range or replaced.

Flame detectors, fire-gas detectors, and other fire detectors should be tested for operation in accordance with instructions supplied by the manufacturer or other test methods acceptable to the local authority.

Inspection forms are also issued for automatic fire detectors in addition to system certification.

The inspection should include the following information on initial tests:

1. Date.
2. Name of property.
3. Address.
4. Installer/maintenance company name, address, and representative.
5. Approving agency(ies) name, address, and representative.
6. Number and type of detectors per zone for each zone.
7. Functional test of detectors.
8. Check of all smoke detectors.

9. Loop resistance for all fixed temperature line-type detectors, if any.
10. Other tests as required by equipment manufacturers.
11. Signature of tester and approval authority representative.

Periodic Testing and Maintenance

Maintenance: Proper maintenance is as important as regular testing in a fire alarm signaling system. Automatic fire detector maintenance depends on the specific type of detector used, local environmental conditions, and manufacturer's recommendations. The manufacturer's recommendations should be implemented in the maintenance program to maintain system reliability.

Periodic Testing of Automatic Fire Detectors: Each detector should be visually inspected to ensure that it remains in good physical condition and that there are no changes such as building modifications, occupancy hazards, and environmental effects that would affect detector performance.

Detectors require periodic cleaning to remove dust or dirt that has accumulated. The frequency of cleaning depends upon the type of detector and the local ambient conditions. For each detector, the cleaning, checking, operating, and sensitivity adjustment should be performed according to the manufacturer's instructions. These instructions should detail methods such as vacuuming to remove loose dust and insects, and washing to remove heavy greasy and grimy deposits. The manufacturer may provide cleaning service at the factory or at field locations in lieu of these cleaning methods.

All automatic detectors suspected of exposure to a fire condition should be tested. Inspection forms should be issued each time a periodic test and inspection is performed. The form is identical to the initial inspection form except that it includes information on test frequency; name of person performing inspection, maintenance and/or tests, affiliation, business address, and telephone number; and designation of the detector(s) tested.

Smoke, Flame, and Other Fire Detectors: All smoke detectors should be visually inspected in place at least semiannually to identify missing detectors, detectors with impeded smoke entry, abnormally dirty detectors, and detectors no longer suitably located because of occupancy or structural changes. Smoke detectors should be tested annually.

Smoke detectors need to be cleaned for optimum detection results. If dust and dirt partially surround the smoke chamber of an ionization detector, smoke particulate may not reach the chamber and the detector becomes less sensitive. If dust and debris insulate the radioactive foil, the rate of ionized air is reduced, making the ionization detector more sensitive and prone to false alarm. Accumulation of dust and film on photoelectric sensors such as on bulbs, lenses, and photocells will diminish the intensity of light within the detector. Light-scattering-type detectors will become less sensitive as the light intensity is decreased. Light-attenuation-type detectors will become more sensitive as the light intensity is decreased.

Detector sensitivity should be checked within one year after installation and every alternate year thereafter. Detectors that are abnormally sensitive should be replaced or cleaned and recalibrated.

Air duct detectors should undergo additional tests, consisting of visual inspection of the detector installation (including seals) to detect any abuse or modification of the device or installation and its intended operation. Also, the manufacturer's recommendations should be used to verify that the device will respond to smoke in the air stream (e.g., measuring pressure drop or air flow through the detector for devices using sampling tubes is acceptable).

Flame detector sensitivity is affected by dust and dirt built up on the lens. The lens should be periodically cleaned for optimum detector performance; all flame detectors should be maintained in a clean condition for optimum detection results.

All flame detectors, fire-gas detectors, and other fire detectors should be tested at least semiannually as prescribed by the manufacturer, and more often if necessary.

Heat Detectors: For nonrestorable spot-type detectors, after the 15th year, at least two detectors out of every hundred (or fraction thereof) should be removed every five years, sent to a testing laboratory for testing, and replaced with new detectors. If a failure occurs on any of the detectors removed, additional detectors should be removed and tested until either a general problem involving faulty detectors or a localized problem involving one or two defective detectors is found.

For restorable heat detectors (except the pneumatic line type), one or more detectors on each signal initiating circuit should be tested at least semiannually and different detectors selected for each test. Each detector should have been tested within five years.

All pneumatic line-type detectors should be tested for leaks and proper operation at least semiannually. Nonrestorable line-type fixed temperature detectors should also be tested for alarm function at least semiannually. The loop resistance should be measured, recorded, and compared with that previously recorded and any change in loop resistance investigated.

Fire Alarm Signaling Systems: In general, a fire alarm signaling system should be properly tested and inspected at prescribed intervals. For most systems, the local authority should be consulted on all system alterations and additions. The owner of the protected facility is responsible for providing proper system maintenance by either personnel at the facility or through a maintenance agreement with outside specialists.

All apparatus should be restored to normal as promptly as possible after each test or alarm, and kept in normal condition for operation. Apparatus should be rewound, reset, or replaced as necessary. A complete record of the tests and operations of each system should be kept, in general, for at least one year. The record should be available for examination and reported to the local authority when required.

Central Station Systems: The central station should have a minimum of two persons on duty to ensure immediate attention to signals received. To provide prompt runner service to the protected premises, one of the two persons may leave the central station if provisions are made to contact the central station at least once every 30 minutes. This may be modified by the local authority when alarms are automatically transmitted to the public fire service communication center. Operation and supervision should be the primary functions of the operators and runners.

Necessary repairs to a central station should be commenced by competent maintenance personnel within four hours after notification of a need for service and continue until completion. The central station should maintain a suitable stock of spare parts to allow updating and repair of equipment at the central station and the protected premises.

Tests of all circuits extending from the central station, and tests of central station devices, should be made at intervals not to exceed 24 hours. Complete and satisfactory tests should be performed monthly on all systems and devices except as follows:

1. Quarterly or more frequently for all transmitters, waterflow actuated devices, automatic fire detection systems, and valve supervisory devices. After any vertically mounted sprinkler system control valve has been operated, the owner or occupant should be encouraged to perform a drain test to ensure that the valve has been fully reopened.

2. Annually for manual fire alarm boxes, combination night guard and fire alarm boxes, tank water level devices, building and tank water temperature supervisory devices, and other sprinkler system supervisory devices.

3. Automatic fire detection devices on the system should be inspected and tested in accordance with NFPA 72, *National Fire Alarm Code*.

In very large central station facilities the test program should be arranged to maintain maximum protection and service at all times.

Periodic testing is also necessary for the protected premises systems, which are connected to the central station.

Plant management and the central station should be notified just prior to testing and the name of the central station operator recorded. Each waterflow alarm device should be actuated by use of the inspector's test facilities. (Ringing of the local alarm bell is not an indication of transmitter operation.)

For sprinkler waterflow alarm tests, an actual waterflow, through the use of a test connection, should be used to test reliability of the sprinkler alarm unit as a whole. For a wet-pipe system, the test connection at the extremity of the system should be used.

Automatic fire detectors should be tested. Other alarm transmitters and supervisory switches and transmitters should be actuated in accordance with their frequency schedule. Circuits within the protected premises in ground and open conditions should be tested for proper operation. Primary power should be turned off and one fire alarm transmitter operated from batteries. The system, and primary power, should subsequently be restored to normal operating conditions. The central station and plant management should be notified of test conclusion and the number and identity of signals checked. When several waterflow switches are connected to a single transmitter and test jacks are provided, the first and last tests should be made without the use of the test jack. For periodic testing of multiplex systems, the above procedure applies; "transponders" replace "transmitters" as described in the procedure.

Local System Testing: Periodic testing and inspection of local systems at prescribed intervals, in general, should be performed at the discretion of the qualified persons in charge of the system.

Auxiliary Systems: The auxiliary system should be inspected during monthly tests to observe the condition of all components and to determine any changes in the property that may affect the protection. Reports of test results and changes should be furnished to the local authority. All initiating and transmitting devices on the system, including transmission of signals to the municipal communication center, should be tested monthly. (The local authority should be notified whenever an auxiliary system is not in service.)

Noncoded manual fire alarm boxes should be tested at least once every year and fire detectors tested according to NFPA 72, *National Fire Alarm Code*.

If an engine-driven generator dedicated to the protective signaling system is used as a required power source it should be tested by being operated weekly under load, disconnecting the normal supply to the system for a minimum of one-half hour in a continuous period.

Remote Station and Proprietary Systems: All operator controls at the remote station designated by the authority having jurisdiction should be tested at each change of shift, and drills conducted at regular intervals to satisfy the local authority.

Tests should be conducted annually for all automatic fire detection systems or other systems and devices except for waterflow-actuated devices and other equipment.

Waterflow-actuated devices should be tested at least every quarter. For sprinkler waterflow alarm tests, an actual waterflow, through the use of a test connection, should be used to test the reliability of the sprinkler alarm unit as a whole. Gate valve supervisory switches, manual fire alarm boxes, combination guard tour and fire alarm boxes, tank water level devices, building and tank water temperature supervisory devices, and other sprinkler system supervisory devices should be tested quarterly.

After any vertically mounted sprinkler system control valve has been operated, a drain test should be performed to ensure that the valve has been fully reopened. If possible, one of the supervisory switches tested at each inspection should be the most electrically remote device on the circuit being tested.

A flow through the alarm test bypass connection should be used to test the waterflow alarm of a dry-pipe, pre-action, or deluge sprinkler system. For a wet-pipe sprinkler system, the test connection at the extremity of the system should be used.

Testing the waterflow alarm device for dry-pipe, deluge, or pre-action systems, using the bypass test valve, does not require tripping of the dry-pipe, deluge, or pre-action valve when all related equipment is maintained in proper operating condition. Trip tests of dry-pipe, deluge, or pre-action valves are usually the responsibility of the property owner or lessee.

The test point for a wet sprinkler system should have an orifice sized for the smallest sprinkler head in the system and be located in accordance with NFPA 13, *Standard for the Installation of Sprinkler Systems*.

If an engine-driven generator dedicated to the protective signaling system is used as a required power source, it should be tested by being operated weekly under load by disconnecting the normal supply to the system for a minimum of one-half hour in a continuous period.

Emergency Voice/Alarm Systems: Emergency voice/alarm communication systems should be maintained and tested by qualified personnel. Functional and operational testing of the voice/alarm signaling service should be conducted annually and include the usage of a representative number of reporting devices, i.e., two-way telephones for the fire service, the fire warden, or general public emergency use, and the automatic and manual voice paging systems in each zone. All elements of the system should be tested at least annually.

Notification Appliances: Notification appliances for a protective signaling system should be tested at the same interval. The notification appliances for the private operating mode should be tested once every six months. Notification appliances may require periodic cleaning to remove dust or dirt that has accumulated. The frequency of cleaning depends on the local ambient conditions. For each appliance, the cleaning, checking, operation, and volume adjustments should be attempted only after consulting the manufacturer's instructions.

A permanent record showing all details of the test, including the name of the inspector, the type number, the location, and the date, should be kept on the premises for at least five years.

System Identification Placard: Periodic testing verifies the ongoing operation of the system. A complete functional test of a protective signaling system includes testing of connections to any equipment monitored or controlled by the system. A permanently mounted identification placard should be located in or adjacent to the protective system control unit and contain the following information:

1. Names, addresses, and telephone numbers of the installation and servicing contractors.
2. Reference to the standard (including date of issue) to which the system conforms.
3. Description of primary (main) and secondary (standby) power supplies. This should include physical location and identification of all overcurrent devices and control switches, type of secondary supply, standby battery specifications, standby generator, names of local inspection authorities (including references and dates), and location of the as-built drawings and system operating and maintenance instructions.

Placard information should be verified in the acceptance procedure as previously described in this chapter.

Periodic tests should be performed in accordance with recommended schedules, or more frequently when required by the local authority. When less than a 100 percent test is being performed, a record should be maintained of the individual devices being tested so that different devices and appliances are used in subsequent tests.

Public Fire Service Communication Systems: Test procedures for an auxiliarized system cover only testing of the alarm equipment up to and including the trip coil in the master box. Additional test procedures are needed for the balance of the system; i.e., the master box through to, and including, the public fire service communication center.

Testing facilities should be installed at the communication center and each satellite communication center (if used). Those facilities for systems leased from a nonmunicipal organization may be located elsewhere if approved by the local authority.

An emergency power source other than batteries should be operated to supply the system for a continuous period of 1 hr at least weekly. This test should require simulated failure of the normal power source.

Although public fire service communication systems are usually tested and maintained by municipal employees, this service can be performed by private contractors. In this case, complete written records of the installation, maintenance, test, and extension of the system should be forwarded to the responsible municipal employee as soon as possible. The municipal employee should also be notified of any failure and restoration of service.

Maintenance performed by other than the municipality or a municipal employee should be under written contract, guaranteeing performance acceptable to the local authority.

Telephone receiving equipment, i.e., the power source, incoming circuits, and public safety answering point (PSAP) equipment, should also be tested. Circuits should be tested at a minimum of once a week, and PSAP equipment operated in the failsafe mode at least once a week. All dispatch system apparatus should be restored to normal condition as promptly as possible after each test or alarm in which the apparatus functioned.

Where supervisory devices or tests indicate that trouble has occurred anywhere on the system, the operator should take appropriate steps to repair the fault or, if this is not possible, isolate the fault and notify the official responsible for maintenance.

Manual test of dispatch circuit instruments must be made and recorded at least once in each 24-hr period. Circuits for graphic transmission of signals should be tested by a message transmission. Outside devices, radio, telephone, or other facilities for alerting volunteer and off-duty fire fighters should be tested daily, and all wired radio and voice amplification circuits subjected to a talking test at least twice daily.

In the communications center, manual tests of the power supply for wired dispatch circuits should be made and recorded at least once in each 24-hr period. Such tests should include:

1. Current strength of each circuit and voltage across terminals of each circuit, inside protected devices. Changes in voltage or current of any circuit that amount to 10 percent of normal should be investigated immediately.

2. Voltage between ground and circuits. When this test shows a reading in excess of 50 percent of that shown in the voltage test, the trouble should be immediately located and cleared. Readings in excess of 25 percent should be checked. These readings should be taken with a voltmeter of not more than 100 ohms resistance per volt to minimize false ground readings.

   NOTE: Systems in which each circuit is supplied by an independent current source will require tests between ground and each side of each circuit. Common-current source systems will require voltage tests between ground and each terminal of each battery and other current source.

3. A ground current reading is acceptable in lieu of item 2. When this method of testing is used, all grounds showing a current reading in excess of 5 percent of the normal line current should be checked immediately.

4. Voltage across terminals of common battery, on switchboard side of fuses.

5. Voltage between common battery terminals and ground. Abnormal ground readings should be investigated immediately. If more than one common battery is used, each should be tested.

Dispatch circuit instruments should be manually tested at least once each 24-hr period.

For equipment in the fire station, the power supply for dispatch circuits should be manually tested and recorded at least once in each 30 days. Such tests should include the voltage across terminals of any power source on the receiving device side of fuses, and the voltage between any power source terminals and ground. Abnormal ground readings should be investigated immediately. Emergency generator equipment should be operated to supply the system for a continuous period of 1 hr at least weekly. This test should also require simulated failure of a normal power source. Batteries supplying dispatch equipment should be tested as described in Chapter 8, "Power Supplies."

Public reporting systems should also be tested and maintained. A complete record should be kept by the municipality of all test and alarm signals, all circuit interruptions and observations or reports of apparatus failure or derangements, and all seriously abnormal or defective circuit conditions indicated by test or inspection for public reporting systems. These records should include the date and time of all occurrences.

Tests should be conducted from the multiplex interface to verify trouble indications for common-mode failures, such as alternating current power failure. To test for the alternating current power, the alternating current power supply to the interface should be removed and at least one initiating device activated. The manufacturer's manual should be consulted for other common-mode failures, and the described testing procedures conducted. Proper receipt and proper processing at the central supervising station should be verified.

Each fault condition the system is required to detect should be introduced on the signaling line circuit. Proper receipt and the proper processing of the signal at the central supervising station should be verified.

When multiplex protective signaling systems are equipped with various optional features unique to these systems, the manufacturer's manual should be consulted to determine the proper testing procedures for verifying the proper operation of these optional features.

Manual tests of box circuit instruments should be made and recorded at least once each 24 hours. Where applicable, all box circuit instruments should be tested by use of operators' keys. Where repeating facilities are necessary, the test of one box from every circuit from which no alarm was transmitted during the past month should be transmitted over the entire system. Boxes should be tested by operation under conditions simulating actual use. Test signals should be transmitted and recorded at the public fire service communication center. A periodic test should be performed on all fire alarm boxes at least once in each 60-day period, and the boxes examined, cleaned, and all functions tested.

Records of boxes should include: box identification, location address, circuit number (if applicable), physical mounting, description by manufacturer, model number, date of installation and power source (radio), and test dates and time. Field inspection forms for the boxes should include information on physical condition, paint, mounting, door function, drop wire or antenna; tests of all box functions; and maintenance.

Each coded radio box should automatically transmit a "test" message at least once in each 24-hr period. The test should include the operation of all message functions associated with each box tested. Such message functions should be transmitted to the respective communication center, received, and permanently recorded. Where solar charging of box battery-(ies) is utilized, the solar cell associated with each box in the system should be examined and cleaned at least once in each 60-day period. Receiving equipment associated with coded radio-type systems should be tested at least once each hour. The receipt of test messages is considered sufficient, provided at least one such message is received each hour.

Manual tests of box and dispatch circuit instruments for telephone reporting systems should be performed and recorded at least once each 24-hr period. If a telephone (parallel) reporting system is used, the person testing a voice box should furnish identification and request the calling location.

RECORDS

Complete records, sufficient to ensure reliable operation of all alarm system functions, should be maintained in a satisfactory manner.

A complete record should be kept by the municipality of all test and alarm signals, all circuit interruptions and observations or reports of apparatus failures or derangements, and all seriously abnormal or defective circuit conditions indicated by test or inspections. These records should include the date and time of all occurrences.

When a combination of leased/owned facilities exists, records that must be maintained by the lessor for the municipality should be specified. A report of operations summarizing important statistics should be prepared annually. Records of wired circuits (box and dispatch) should include outline plans showing terminals and box sequence; diagrams of office wiring; and materials including trade name, manufacturer, and year of purchase or installation. Emergency generating equipment periodic test records should include date and time; fuel, electrical, coolant, and exhaust system conditions; and operating time.

MULTIPLEX SYSTEMS AND CIRCUIT STYLES

Additional acceptance and re-acceptance test methods are needed to verify multiplex-type protective signaling system performance. The manufacturer's manual and the as-built drawings provided by the supplier should be used to verify proper operation after the initial testing phase has been performed by the supplier or a designated representative.

Starting from the unpowered condition, the system should be initialized in accordance with the manufacturer's manual. One or more tests should be conducted to verify that communication exists between the central processing unit and the connected central supervisory station peripheral devices.

Each initiating device circuit should be tested for its alarm reporting capability by operating at least one of the initiating devices connected to it. Upon completion of this test, an open-circuit trouble condition should be made to verify open-circuit fault detection. Additional tests should be conducted to verify all possible status modes that the initiating device circuit should provide. Proper receipt and processing of each test signal by the central supervisory station equipment should be verified.

When the style of an initiating device circuit or signaling line circuit requires alarm signaling capability in the presence of a fault condition, that capability should be verified by the proper receipt and the proper processing of the alarm signal at the central supervising station. This test should be repeated for all fault conditions for the style of the circuit involved.

Table 14-3. Testing Frequencies

	Initial & Reacceptance Tests	Monthly	Quarterly	Semiannually	Annually
1. Alarm Notification Appliances					
a. Audible Devices	X				X
b. Speakers	X				X
c. Visible Devices	X				X
2. Batteries—Central Station Facilities					
a. Lead-Acid Type					
1. Charger Test	X				X
(Replace battery as needed.)					
2. Discharge Test (30 min.)	X	X			
3. Load Voltage Test	X	X			
4. Specific Gravity	X			X	
b. Nickel-Cadmium Type					
1. Charger Test	X		X		
(Replace battery as needed.)					
2. Discharge Test (30 min.)	X				X
3. Load Voltage Test	X				X
c. Sealed Lead-Acid Type	X	X			
1. Charger Test		X	X		
(Replace battery as needed.)					
2. Discharge Test (30 min.)	X	X			
3. Load Voltage Test	X	X			
3. Batteries—Fire Alarm Systems					
a. Lead-Acid Type					
1. Charger Test	X				X
(Replace battery as needed.)					
2. Discharge Test (30 min.)	X			X	
3. Load Voltage Test	X			X	
4. Specific Gravity	X			X	
b. Nickel-Cadmium Type					
1. Charger Test	X				X
(Replace battery as needed.)					
2. Discharge Test (30 min.)	X				X
3. Load Voltage Test	X			X	
c. Primary Type (Dry Cell)					
1. Load Voltage Test	X	X			
d. Sealed Lead-Acid Type					
1. Charger Test	X				X
(Replace battery every 4 years.)					
2. Discharge Test (30 min.)	X				X
3. Load Voltage Test	X			X	

NOTE: For testing addressable and analog described devices, which are normally affixed to either a single molded assembly or twist lock type affixed to a base, testing must be done utilizing the signaling style circuits (Styles 0.5 through 7). Analog type detectors shall be tested with the same criteria.

Table 14-3. Testing Frequencies (continued)

	Initial & Reacceptance Tests	Monthly	Quarterly	Semiannually	Annually
4. Batteries—Public Fire Alarm Reporting Systems	X (daily)				
a. Lead-Acid Type					
1. Charger Test	X				X
(Replace battery as needed.)					
2. Discharge Test (2 hours)	X		X		
3. Load Voltage Test	X		X		
4. Specific Gravity	X			X	
b. Nickel-Cadmium Type					
1. Charger Test	X				X
(Replace battery as needed.)					
2. Discharge Test (2 hours)	X				X
3. Load Voltage Test	X		X		
c. Sealed Lead-Acid Type					
1. Charger Test	X				X
(Replace battery as needed.)					
2. Discharge Test (2 hours)	X				X
3. Load Voltage Test	X		X		
5. Conductors/Metallic	X				
6. Conductors/Nonmetallic	X				
7. Control Equipment: Fire Alarm Systems Monitored for Alarm, Supervisory, Trouble Signals					
a. Functions	X				X
b. Fuses	X				X
c. Interfaced Equipment	X				X
d. Lamps and LEDs	X				X
e. Primary (Main) Power Supply	X				X
f. Transponders	X				X
8. Control Equipment: Fire Alarm Systems Unmonitored for Alarm, Supervisory, Trouble Signals					
a. Functions	X		X		
b. Fuses	X		X		
c. Interfaced Equipment	X		X		
d. Lamps and LEDs	X		X		
e. Primary (Main) Supply	X		X		
f. Transponders	X		X		
9. Control Unit Trouble Signals	X				X
10. Emergency Voice/Alarm Communication Equipment	X				X

NOTE: For testing addressable and analog described devices, which are normally affixed to either a single molded assembly or twist lock type affixed to a base, testing must be done utilizing the signaling style circuits (Styles 0.5 through 7). Analog type detectors shall be tested with the same criteria.

Table 14-3. Testing Frequencies (continued)

	Initial & Reacceptance Tests	Monthly	Quarterly	Semiannually	Annually
11. Engine-Driven Generator	X (weekly)				
12. Fiber Optic Cable Power	X				X
13. Guard's Tour Equipment	X				X
14. Initiating Devices					
a. Duct Detectors	X				X
b. Electromechanical Releasing Devices	X				X
c. Extinguishing System Switches	X				X
d. Fire-Gas and Other Detectors	X				X
e. Heat Detectors	X				X
f. Fire Alarm Boxes	X				X
g. Radiant Energy Fire Detectors	X				X
h. Smoke Detectors - Functional	X				X
i. Smoke Detectors - Sensitivity					
j. Supervisory Signal Devices	X		X		
k. Waterflow Devices	X		X		
15. Interface Equipment	X				X
16. Off-Premises Transmission Equipment	X		X		
17. Remote Annunciators	X				X
18. Retransmission Equipment	X				
19. Special Hazard Equipment	X				X
20. Special Procedures	X				X
21. System and Receiving Equipment—Off-Premises					
a. Operational					
1. Functional—All	X				X
2. Transmitters—WF & Supervisory	X		X		
3. Transmitters—All Others	X				X
4. Receivers	X	X			
b. Standby Loading—All Receivers	X	X			
c. Standby Power					
1. Receivers—All	X	X			
2. Transmitters—All	X				X
d. Telephone Line—All Receivers	X	X			
e. Telephone Line—All Transmitters	X				X

NOTE: For testing addressable and analog described devices, which are normally affixed to either a single molded assembly or twist lock type affixed to a base, testing must be done utilizing the signaling style circuits (Styles 0.5 through 7). Analog type detectors shall be tested with the same criteria.

General inspection	Yes	No
1. Roof:		
Is roof covering noncombustible?	——	——
Are scuppers and drains unobstructed?	——	——
Are lightning arrestors in good condition?	——	——
Are skylights protected by screens?	——	——
Is access to fire escapes unobstructed?	——	——
Do fire escape stairs appear to be in good condition?	——	——
Are fire escape stairs unobstructed?	——	——
Are standpipe and sprinkler roof tanks and supports in good condition?	——	——
Are standpipe and sprinkler control valves secured in proper position?	——	——
2. All floors (inspect from top floor to basement):		
Are self-closing fire doors unobstructed and properly equipped with closing devices?	——	——
Are fire exits and directional signs properly illuminated?	——	——
Is emergency lighting system operable?	——	——
Are corridors and stairways unobstructed?	——	——
Are fire exits unlocked and unobstructed?	——	——
Are sprinklers unobstructed?		
Are standpipe hose outlets properly marked and unobstructed?	——	——
Are sprinkler control valves properly labeled and unobstructed?	——	——
Are recorded weekly inspections made of all sprinkler control valves to make certain they are open?	——	——
Are dry-pipe valves (for sprinklers in areas exposed to freezing) in service, with air pressure normal?	——	——
Are all fire-detection and fire-supression systems in service and tested regularly?	——	——
Are sufficient fire extinguishers present?	——	——
Are extinguishers of the proper type? (See NFPA 10, *Standard for Portable Fire Extinguishers.*)	——	——

Figure 14-6.   Firesafety self-inspection form.

Are extinguishers properly hung
and labeled? ____ ____

Are smoking regulations enforced with
employees, and visitors? ____ ____

Are aisles to exit routes unobstructed
and visible? ____ ____

Is housekeeping properly maintained? ____ ____

Are cleaning supplies safely stored?

Are all trash receptacles emptied at
least daily? ____ ____

3. Ground floor:

Do entrance and exit doors provide
unobstructed egress? ____ ____

Is safe egress uncompromised by
security measures? ____ ____

Exterior Inspection

1. Evacuation:

Do all exits, emergency exits, and
fire escapes have unobstructed passage
to safe areas? ____ ____

2. Environment:

Are grounds clear of accumulations of
flammable material? ____ ____

Have neighboring occupancies minimized
exterior fire hazards? ____ ____

Is fire service access clear? ____ ____

Are standpipe and sprinkler system siamese
connections unobstructed and operable? ____ ____

Are hydrants unobstructed? ____ ____

Personnel Inspection

1. Training:

Do all staff members know how to transmit
a fire alarm? ____ ____

Do all staff members know their assigned
duty in evacuating the building? ____ ____

Do all staff members know how and when
to use portable fire extinguishers? ____ ____

Do all staff members know their
responsibilities in fire prevention? ____ ____

Figure 14-6. Continued.

TESTING SCHEDULES

Table 14-3 lists acceptance and periodic tests unless otherwise noted. The recommended test means are not intended to exclude equivalent means, such as self-diagnostic testing. Tests conducted on equipment and wiring in hazardous locations can require special testing procedures.

CONCLUSION

While testing and maintenance are important to a fire alarm signaling system and its components, it is equally important to periodically inspect an entire protected facility. Figure 14-6 shows some of the typical firesafety items that should be checked on a regular basis.

BIBLIOGRAPHY

NFPA Codes, Standards, Recommended Practices, and Manuals. (See the latest *NFPA Catalog* for availability of current editions of the following documents.)

NFPA 10, *Standard for Portable Fire Extinguishers.*

NFPA 13, *Standard for the Installation of Sprinkler Systems.*

NFPA 70, *National Electrical Code.*

NFPA 72, *National Fire Alarm Code.*

NFPA 101, *Life Safety Code.*

NFPA 1221, *Standard for the Installation, Maintenance, and Use of Public Fire Service Communication Systems.*

References Cited

1. "Study on Central Stations Effectiveness," Fire Chiefs of Los Angeles County, Dec. 30, 1983.
2. "Hotel Fire Alarm Life Safety Systems Audit," J. A. Drouin, 1983.
3. ANSI S.1.4., *Specification for Sound Level Meters,* American National Standards Institute, New York, NY, 1983.

Additional Reading

NBS, *"An Instrument to Evaluate Installed Smoke Detectors,"* NBSIR 78-1430, National Bureau of Standards, Washington, DC, 1978.

UL 217, *Standard for Safety Single and Multiple Station Smoke Detectors,* Underwriters Laboratories Inc., Northbrook, IL, 1993.

UL 268, *Standard for Safety Smoke Detectors for Fire Protective Signaling Systems,* Underwriters Laboratories Inc., Northbrook, IL, 1989.

15

Public Fire Service
Communication Systems

INTRODUCTION

The public fire service communication center receives notification of a fire condition that was initiated by any fire alarm signaling system — local, central station, auxiliary, remote station, proprietary, and household warning systems. Notification must be direct, via an auxiliary or remote station system, or by phone or other means of retransmission. The communication center must be properly maintained and operated so the link is maintained between the fire department, alarm system, and other points within the communication system.

This chapter describes installation, maintenance, and use of a public fire service communication system (and the facility in which it is located); and communication center operation, including retransmission of alarms and reporting and dispatch systems. For additional information see Chapter 3, "Signal Initiation"; Chapter 4, "Signal Transmission"; Chapter 5, "Signal Processing"; and Chapter 14, "Testing."

Public fire service communication systems and facilities include (but are not limited to) public reporting, dispatching, telephone, and both two-way and microwave radio systems. All of these fulfill two principal functions: (1) receipt of fire alarms or other emergency calls from the public and (2) retransmission of these alarms and emergency calls to fire companies and other interested agencies.

A public fire service communication center is defined as the building or portion of a building that houses the central operating part of the fire alarm system. The center is also usually the place where the necessary testing, switching, receiving, retransmitting, and power supply devices are located.

COMMUNICATION CENTER LOCATION AND CONSTRUCTION

If the fire service communication center building is located within 150 ft (46 m) of another structure, special attention should be given to guard against damage by protecting openings and constructing the roof to resist damage that might be caused by falling walls. The communication center should not be located below grade unless the structure is specifically designed for such a location. Floor elevation should be above the 100-year flood plan prediction if the location is below grade.

Seismic and wind loads prevalent in the geographic location of the building should be considered in design and construction of the communication center. Applicable building codes should always be followed. The building housing the center should be of fire resistive construction, or protected noncombustible/limited combustible construction.

If the building is unprotected noncombustible/limited-combustible or ordinary construction, it should have a Class A fire resistive roof covering and a sprinkler system in all areas of the building except the communication center and power room. The sprinkler system should be completely supervised by the communication center.

A fire service communication center can be located in other than a fire department building. In these buildings, the communication center should be separated from the other portions of the building by vertical and horizontal separations with minimum 2-hr fire resistance rating. Openings should be protected by self-closing or automatic fire doors or other assemblies with a minimum $1\frac{1}{2}$ hr fire resistance rating. If spaces adjoining the communication center are ordinary-hazard occupancies (as defined in NFPA 13, *Standard for the Installation of Sprinkler Systems*), they should have an automatic fire alarm system. If such spaces are occupied by extra-hazard occupancies as defined therein, they should have an automatic sprinkler system. Interior finish material should have a maximum flame spread rating of 25.

The communication center and other buildings that house essential operating equipment should be secure. Entryways leading directly from the exterior should be protected by two doors and a vestibule. Entry to the communication center should be restricted to authorized persons only, and door openings protected by not less than a Class B self-closing fire door assembly.

Warm air heating, ventilating, and air conditioning services should be provided to the communication center by independent systems. Main water, sewer, storm sewer, or sprinkler lines should not pass through the communication center or its equipment rooms. An automatic fire alarm system connected to an audible and visible warning device at a continually manned location should protect the entire communication center.

Fire extinguishers should be provided for the communication center with at least two extinguishers with 2-A or greater rating and two extinguishers with a combined rating of 20-B:C or greater. If it is not possible to have these, two multipurpose extinguishers with 2-A:10-B:C rating may be provided.

The communication center should also be equipped with an emergency lighting system that can be immediately placed in service and powered by an independent source. Illumination should be sufficient to permit all necessary operations. At least one self-charging, battery-pack lantern should be available that lights automatically when power is interrupted.

Two sources of power should be provided to operate the communications network and its supporting related systems and equipment under all conditions. Three recommended power sources for the communication center that are considered acceptable are:

1. One circuit from a utility distribution system and a second from an engine-driven generator plus a 4-hr capacity standby storage battery.
2. Two circuits from separate utility distribution systems, serviced or connected so that normal supply to one will not be affected by trouble in the other. This requires supply from two building services on entirely separate distribution networks from independent generating stations.
3. Two engine-driven generators with one unit supplying normal system power and a standby unit that activates within 30 seconds. A standby storage battery having a 4-hr capacity should be provided.

All standby storage batteries incorporated into a power source network should be equipped with suitable float or trickle chargers. If two engine-driven generators are used as a second power source, the fire alarm system 4-hr battery can be omitted.

OPERATION OF THE COMMUNICATION CENTER

Reports of fires and other emergencies to the fire service originate from three principal sources: (1) the general public, (2) the business community (industrial, institutional, commercial and mercantile), and (3) other public service/safety agencies. The reporting process can be made through any of the following means (individually or in combination):

public commercial telephone systems, public emergency reporting/fire alarm systems, privately operated automatic alarm systems, or two-way radio communication systems.

Commercial Telephone Facilities

The conventional commercial telephone network is the most common method of reporting a fire emergency. The installation of outdoor telephone booths increases the availability of telephones for emergency reporting.

The "enhanced 911" system automatically provides both the calling number and the location of the related telephone instrument directly to the dispatcher. This indication is presented visually on a cathode ray tube (CRT), and can be automatically recorded if the communication center includes computer-aided dispatch (CAD) or computerized logging support systems. The expanding implementation of "911" systems has eliminated the necessity of public reporting stations (street boxes), at least in those areas where publicly accessible telephones are well distributed and functional.

Although there are some disadvantages to total dependency upon telephones, they are often the only means of communication available in cities, suburbs, and rural areas. The public telephone system, in general, is a good means to report fires and other emergencies.

Municipal Fire Alarm Systems

Any municipal fire alarm system, whether a coded, voice, or code-voice alarm system, must provide a means by which an alarm can be transmitted from a street alarm box to the communication center.

Use of a public emergency reporting station (fire alarm box) eliminates the difficulty of the dispatcher (who receives the alarm) in determining the location from where the alarm is being transmitted. When actuated, each device should transmit a distinct numerical code in addition to any other functions or capabilities provided.

A Type A public reporting system is one in which an alarm from a fire alarm box is received and is retransmitted to fire stations either manually or automatically. (See Figure 15-1.) A Type B is one in which an alarm from a fire alarm box is automatically transmitted to fire stations and, if used, to outside alerting devices. (See Figure 15-2.)

The Type A system is usually permissible in any size municipality or area and should be provided when there are more than 2,500 emergency calls from boxes per year, or where more than 2,500 alarms are transmitted over the dispatch circuits.

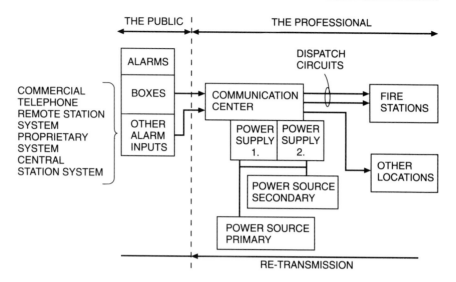

Figure 15-1. Type A public reporting system.

Automatic retransmission of alarms from boxes by use of electronic equipment can be allowed if reliable facilities and override capability are provided. A public emergency reporting system can transmit other emergency signals or calls as long as they do not interfere with the proper handling of fire alarms.

Dispatch Circuits and Equipment

A dispatch circuit is the means by which the fire alarm dispatcher notifies fire companies to respond to an alarm. The location from where the alarm was received is the minimum information that should be transmitted.

Two separate means of transmitting alarms to fire stations should be provided at the communication center. (Only one means of transmission is necessary when fewer than 600 alarms per year are received.) Each alarm transmitted, the date, and the time should be automatically recorded. Devices for transmitting coded or other types of signals should be arranged for manual setting and operation.

Computer-Aided Dispatch (CAD)

Computer-aided dispatch (CAD) is a process by which a computer and its associated terminal(s) is/are used to provide relevant dispatch data (running assignments, address locations, equipment status, utility locations, special hazards, etc.) to the dispatcher or operator concerned.

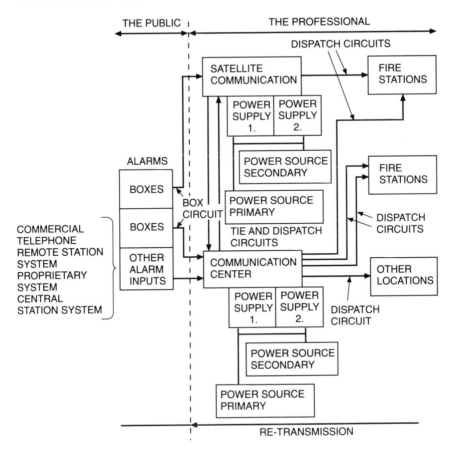

Figure 15-2. Type B public reporting system.

CAD is used to assess related response information, and it identifies companies normally assigned and their current status. (See Figure 15-3.)

The degree to which a CAD system can be implemented to dispatch fire units depends upon the size of the fire department. A small department with one to three stations does not require a complex CAD system with its mobile digital terminals, station terminals, and printers. Small departments can use a personal computer with a dispatch software package to dispatch fire units.

Personal computers should be carefully selected and used and matched to the proper application. The dispatch computer should not be used for other department applications. It should be kept free of time-consuming processes to allow for rapid retrieval of dispatch information. CAD systems for medium to large departments are more complex and must be custom designed for each department.

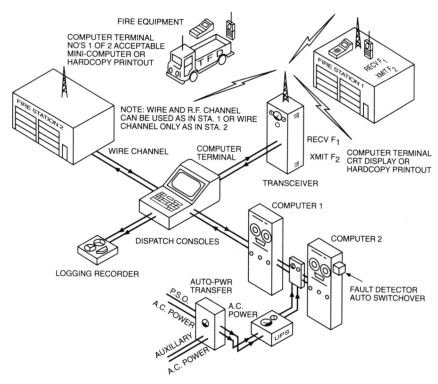

Figure 15-3. Diagram of a CAD system.

There are three major classes of CAD services for fire communication systems. A Class 1 CAD system is one in which computer technology and equipment selects and dispatches fire service personnel and equipment and emergency service assistance. A Class 2 CAD system is used as a support dispatch operation to voice- or graphic-type-operated dispatch systems. A Class 3 CAD system primarily supports fire service dispatching and is limited to status and logging information.

BIBLIOGRAPHY

NFPA Codes, Standards, Recommended Practices, and Manuals. (See the latest *NFPA Catalog* for availability of current editions of the following documents.)

NFPA 10, *Standard for Portable Fire Extinguishers.*

NFPA 13, *Standard for the Installation of Sprinkler Systems.*

NFPA 220, *Standard on Types of Building Construction.*

NFPA 1221, *Standard for the Installation, Maintenance, and Use of Public Fire Service Communication Systems.*

16

Cost Analysis

INTRODUCTION

Economics, i.e., cost, is a prime motivator in determining the amount and type of fire protection that is installed in a facility. This chapter describes various methods of cost analysis that can be used to determine economic feasibility of various fire alarm signaling systems.

In economic analysis, the economic impact of design, construction, and equipment installation is studied to determine the relative worth of net economic gains to be expected from alternative solutions relative to their net economic costs. Resources can be easily wasted if an economic analysis is not performed.

Not all fire protection decisions, however, require a detailed comprehensive economic analysis. Determining whether to place a smoke detector in one small room of a building, for instance, would not justify a detailed analysis.

BASIC PRINCIPLES OF ECONOMIC ANALYSIS

Sound economic analysis is based on an objective, factual study of all possible alternatives, independent of financing. The analysis period should not extend beyond a reasonable forecast period. Since the analysis is a study of the future, past events and investments (or "sunk costs") are irrelevant. When preparing an economic analysis, factors used should have identical time periods, be separated into market versus nonmarket, and be discounted to the same time/date. Common factors of equal magnitude can be omitted from the analysis.

In cost analysis, the process of achieving a rational decision is made by a logical method of analysis. Eight key elements are:

1. Recognition of a problem — the realization that a problem exists.

2. Definition of the goal or objective to be accomplished. What is the task?
3. Assembly of relevant data. What are the facts? Is additional data needed? Is the additional information worth the cost of obtaining it?
4. Identification of feasible alternatives. What are the practical alternative ways of accomplishing the objective or task?
5. Selection of the criteria (e.g., political, economic, ecological) for judging the best alternative.
6. Construction of the various interrelationships; frequently called mathematical modeling.
7. Prediction of outcome for each alternative.
8. Choice of the best alternative to achieve the objective.

The decision process system is not simply proceeding from the first to the last element; in fact, it is often necessary to re-examine earlier elements.

One of the most important concepts of engineering economics that follows the decision-making process is the time value of money, which is a value in the form of the willingness of people to pay interest for the use of money. The economics of fire protection can be estimated correctly only by including a factor (interest) for the economics of money.

DEFINITIONS

Interest: A charge for borrowed money.
Rate of Interest: The percentage (per unit of time) that is paid on money borrowed.
Compound Interest: Received at the end of each time period, this interest is based on the original borrowed amount plus the accumulated interest, using the PFinA or similar formula where

P = a present sum of money
F = a future sum of money
i = interest rate per time period
n = number of time periods
A = uniform periodic (end of period) payment.

LIFE CYCLE COST

The life cycle cost of a fire protection system includes all costs associated with the system during its lifetime. Life cycle costs can include engineering/design costs; costs of initial equipment, installation, and system startup; annual inspection, maintenance, repair, and operation costs; repair costs (other than scheduled annual repair); and salvage cost or value. (See Figure 16-1.)

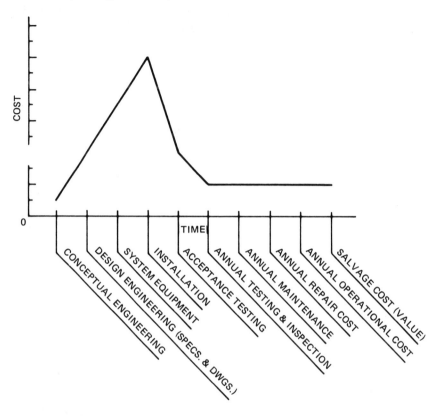

Figure 16-1. Life cycle system curve for fire alarm signaling systems.

Life cycle cost is a method of evaluating expenditures and recognizes the sum total of all costs associated with the expenditure during the time it is in use. Using cost comparisons for economic decision making should include all appropriate costs during the life of a fire protection system.

Alternative fire protection solutions are evaluated by determining the present cost of each alternative and selecting the one at the least cost but most effectiveness.

Two of the easiest ways of comparing alternatives are: (1) present worth of cost analysis and (2) series present worth of cost. Simply stated, the analysis illustrates the maximum present worth of benefits and the minimum present worth of cost. Three criteria for economic efficiency are presented in Table 16-1.

TABLE 16-1. Present Worth Analysis

	Situation	*Criterion*
Fixed input	Amount of money or other input resources is fixed	Maximize present worth of benefits or other outputs
Fixed output	There is a fixed task, benefit, or other output to be accomplished	Minimize present worth of costs or other inputs
Neither input nor output fixed	Neither amount of money or other inputs nor amount of benefits or other outputs is fixed	Maximize (present worth of benefits minus present worth of costs) or, more simply, maximize net present worth

PRESENT WORTH OF COST ANALYSIS

To find the present cost of some future expenditure, the formula for the calculation is:

$$P = \frac{F}{(1 + i)^n}, \text{ or functionally:}$$

where
$P = F \ (P/F, \ i\%,n); \ F = $ future cost, $P = $ present cost, $n = $ number of periods, and $i = $ interest rate per period.

Showing the formula as a time line:

Example: A firm is trying to decide which of two alternative fire alarm signaling systems it should install to replace an outdated system. One system, Alternative A, will cost $110,000 to install in the second year while another system, Alternative B, will cost $125,000 to install in the third year. Assuming all other costs are equal and the cost of borrowed money is 12 percent, which alternative should be selected?

Alternative	Cost	Periods	Interest—%
A	$110,000	2	12
B	$125,000	3	12

Time line for Alternative A is:

$110,000

Time line for Alternative B is:

$125,000

P = present cost.
Present Cost of Alternative A

$$P = \frac{F}{(1+i)^n}$$

$$P = \frac{110,000}{(1+0.12)^2}$$

$$P = \$87,691$$

Present Cost of Alternative B

$$P = \frac{F}{(1+i)^n}$$

$$P = \frac{125,000}{(1+0.12)^3}$$

$$P = \$88,972$$

Since only the differences between the alternatives are relevant, the appropriate economic choice is Alternative A, the system with the minimum present cost.

SERIES PRESENT WORTH OF COST

To find the series present worth of cost of some recurring future sum of money, the formula for the calculation is

$$P = A\frac{(1+i)^n - 1}{i(1+i)^n} \quad \text{or, functionally,} \quad P = A(P/A, i\%, n)$$

where
A = uniform periodic payment, P = present cost, n = number of periods, and i = interest rate per period.

Sketching a time line of this formula shows:

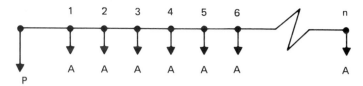

Example: A fire protection engineer working with an architectural/ engineering firm wishes to determine the present cost for annual expenditures to test and maintain fire detectors. The annual cost of functional testing and sensitivity testing for fire detectors is determined to be $1,020 and $3,380, respectively, or $4,400 total. Assuming other annual costs are negligible and the interest on borrowed money is 12 percent, what is the present cost of the annual expenditure if the useful life of the fire detectors is 20 years?

The time line is:

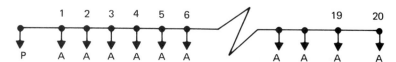

Using the series present cost formula (A = uniform periodic payment, P = present cost, n = 20 yrs, and i = 12%):

$$P = A\left[\frac{(1+i)^n - 1}{i(1+i)^n}\right] = 4,400\left[\frac{(1+0.12)^{20} - 1}{0.12(1+0.12)^{20}}\right]$$

$$P = 4,400\,(7.469)$$

$$P = \$32,864$$

The series present worth of cost factor, i.e.,

$$\frac{(1+i)^n - 1}{i(1+i)^n}$$

does not need to be calculated each time since compound interest tables contain these data. (Interest tables assume that payments are made at the end of each period, so an appropriate adjustment should be made if payments are made at the beginning of each period.)

The following example illustrates the cost analysis for a fire alarm system that is selected based upon the lowest present cost.

Example: A fire protection engineer working with an architectural engineering firm and the facility's owner is trying to decide whether a multiplex fire alarm system or a "hardwired" fire alarm system is the more economical as determined by the life cycle cost of each system. Data for each alternative system are provided in Table 16-2.

TABLE 16-2. Alternative System Data

Cost	Time (yrs)	Alternative 1	Alternative 2
		$	$
Engineering Design	0	15,000	20,000
Equipment	1	110,000	154,600
Installation	1	62,500	41,000
Startup	2	14,000	16,400
Inspection	Annual	12,500	2,000
Testing	Annual	17,400	3,200
Operation & Repair	Annual	1,200	700
Salvage Cost (Value)	20 yrs	(6,500)	(2,000)

The total annual cost of Alternative 1 is $31,100 (total for inspection, testing, operation, and repair).

A time line for Alternative 1 is:

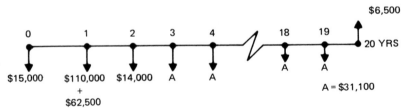

The present cost of Alternative 1 is:

P = 15,000 + 172,500 (P/F, 12%,1) + 14,000 (P/F, 12%,2) + 31,100
(P/A, 12%,17) × (P/F, 12%,3) – 6500 (P/F, 12%,20).

Note that the annual cost payments are made during the end of the second year in which the fire alarm system operates.

By using an interest table, the equation becomes:

P = 15,000 + 172,500 (0.8929) + 14,000 (0.7972) + 31,100 (7.12)
× (0.7118) – 6500 (0.1037).
P = 15,000 + 154,025 + 11,161 + 157,615 – 674 = $337,127.

The present cost of Alternative 1 is $337,127.

For Alternative 2, the total annual cost is $5,900 (total for inspection, testing, operation, and repair).

A = 2,000 + 3,200 + 700 = $5,900.

A time line for Alternative 2 is:

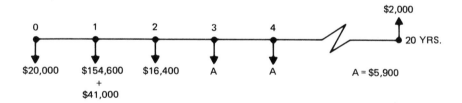

The present cost of Alternative 2 is:

P = 20,000 + 195,600 (P/F,12%,1) + 16,400 (P/F, 12%,2) + 5,900
(P/A,12%, 17) × (P/F,12%,3) – 2,000 (P/F,12%,20).
P = 20,000 + 195,600 (0.8929) + 16,400 (0.7972) + 5,900 (7.12)
× (0.7188) – 2,000 (0.1037).
P = 20,000 + 174,651 + 13,074 + 29,901 – 207.
P = $237,419.

The present cost of Alternative 2 is $237,833.

Alternative 2 should be selected since the criterion of selection is to minimize present cost.

OTHER COST ANALYSIS METHODS

Other techniques can be used to determine the most economic solution of comparable alternatives. These techniques are included herein.

Annual Cash Flow Analysis

Annual cash flow analysis converts money to an equivalent uniform annual cost or benefit. Future sums and present worth of cost are equated to equivalent uniform annual costs.

Example: Find A, given P.

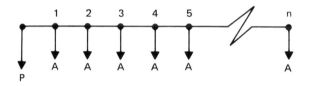

Capital recovery

$$A = P\frac{i(1+i)^n}{(1+i)^n - 1}, \text{ or } (A/P, i\%, n)$$

Example: Find A, given F.

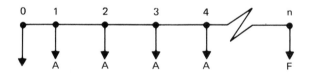

Sinking fund

$$A = F\frac{i}{(1+i)^n - 1}, \text{ or } (A/F, i\%, n)$$

Benefit/Cost Analysis

If the benefits of a project exceed the cost, then economically the project is worth undertaking.

$$\frac{\text{Benefits}}{\text{Cost}} \text{ ratio} = \frac{\text{Present Worth of Benefits}}{\text{Present Worth of Cost}}$$

$$= \frac{\text{Equivalent Uniform Annual Benefit}}{\text{Equivalent Uniform Annual Cost}}$$

This is predicated upon a given minimum attractive rate of return.

Rate of Return: To determine a rate of return on an investment, the various consequences of an investment are converted to a cash flow. Then the cash flow is used to solve for the unknown value, i. The value of i is the rate of return. Two examples of cash flow equations are: (1) present worth of cost = present worth of benefits and (2) equivalent uniform annual cost = equivalent uniform benefits. Once the benefits and costs are known, the unknown rate of return can be calculated.

Payback Period

The payback period is the time required for the profit or other benefits from an investment to equal the cost of the investment; it is also the time required for savings on an investment to equal the cost of the investment.

Payback period calculation is an approximate, rather than an exact, economic analysis calculation. All costs and profits or savings of the investment prior to payback are included without considering differences in timing (when they were acquired). All the economic consequences beyond the payback period are ignored. Because it is an approximate calculation, payback period may or may not select the correct alternative. For example: Both alternatives below have the same initial cost and payback period, and benefits of $3,000 each, so it would seem that either alternative is acceptable. But compare time lines:

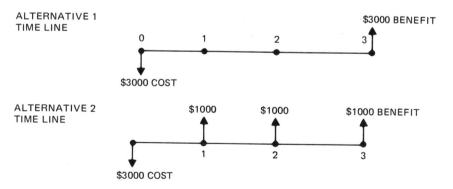

BIBLIOGRAPHY

Additional Reading

Baresh, Norman N., and Kaplan, Seymour, *Economic Analysis for Engineering and Managerial Decision Making*, 2nd ed., McGraw-Hill, Inc., New York, 1978.

Grant, E.L., Ireson, W.G., and Leavenworth, R.S., *Principles of Engineering Economy*, 6th ed., Ronald Press Company, New York, 1976.

Newman, Donald G., *Engineering Economic Analysis*, Engineering Press, San Jose, CA, 1977.

Ostwald, P.F., *Cost Estimating for Engineering and Management*, Prentice-Hall, Inc., Englewood Cliffs, NJ, 1974.

17

Code Requirements

INTRODUCTION

Local code requirements for fire alarm signaling systems can often be confusing, since different codes may all apply to a particular system. The fire alarm signaling system designer must have a thorough understanding of a locality's current code or standard for fire alarm signaling systems and all pertinent requirements. Working knowledge of code requirements is a prime factor in cost-efficient fire alarm signaling system design.

This chapter describes general requirements contained in codes and standards; the role of testing laboratories is briefly described. A tabular listing of some NFPA codes and standards referenced throughout this text is also included. For related information on code requirements, see Chapter 12, "Approvals and Acceptance."

TESTING LABORATORIES

It is important that all equipment in a fire alarm signaling system be listed by an independent testing agency for the appropriate signaling purpose. Use of listed equipment is often mandated by local codes, even though the code may not specify a particular testing agency's listing.

In the listing process, a sample of each fire alarm system component is tested to determine its compliance with one or more of the testing laboratory's standards. Subsequent followup inspections are conducted at the manufacturer's facilities to ensure continued compliance with the test standard. Two major well-known testing laboratories in the United States are Underwriters Laboratories Inc. (UL) and Factory Mutual

Research Corporation (FMRC). UL and FMRC both determine the suitability of fire alarm equipment for its intended service, and test equipment for compliance with appropriate NFPA codes and standards.

Some UL and FMRC test standards that apply to fire alarm signaling systems are shown in Table 17-1.

TABLE 17-1. Testing Laboratory Standards for Fire Alarm Signaling Systems

Underwriters Laboratories Inc. (UL) Standards	
UL 217	*Single and Multiple Station Smoke Detectors*
UL 268	*Smoke Detectors for Fire Protective Signaling Systems*
UL 268A	*Smoke Detectors for Duct Application*
UL 464	*Audible Signals*
UL 521	*Heat Detectors for Fire Protective Signaling Systems*
UL 864	*Control Units for Fire Protective Signaling Systems*
UL 985	*Household Fire Warning System Units*
UL 1480	*Speakers and Amplifiers*
UL 1481	*Power Supplies*
UL 1638	*Visual Signals*
UL 1730	*Annunciator Systems*
Factory Mutual Research Corp. (FM) Standards	
FM 3210	*Thermostats ("Heat Detectors") for Automatic Fire Detection*
FM 3230 through FM 3250	*Smoke-Actuated Detectors for Automatic Fire Alarm Signaling*
FM 3260	*Flame Radiation Detectors for Automatic Fire Alarm Signaling*
FM 3820	*Electrical Utilization Equipment for Fire Alarm Control Equipment*
[no number]	*Facilities and Procedures Audit; Program Manual* (this describes the FM listing process)

Fire alarm signaling system requirements in a particular code or standard depend on the occupancy for which the system is intended. A local occupancy code will specify the type of system and detection that is required. For example: minimum protection for a nursing home could be: (1) a fire alarm system that provides evacuation signals, (2) smoke detectors in building corridors and certain other areas, and (3) system connection to the local fire department. NFPA *101, Life Safety Code*, describes types of fire alarm signaling systems for a variety of occupancies. A summary of the protection requirements in NFPA *101* by occupancy is shown in Table 17-2.

Once the type of system has been determined, installation requirements can be found in NFPA 72, *National Fire Alarm Code*.

REQUIREMENTS IN LOCAL BUILDING AND FIRE CODES

After well-publicized tragic fires in residential and commercial occupancies in the 1970s and 1980s, many cities, counties, and states reviewed their building and fire codes, strengthening the protective aspects of the codes to prevent loss of life and property.

Firesafety codes and standards that are developed by the National Fire Protection Association (NFPA) can be adopted by a local community. The standards can be adopted by reference (title and publishing information on the NFPA standard are mentioned only in the local code) or by transcription (printing of the standard in the local code). Similarly, model building codes are developed by private associations for modification and adoption by local communities. Model building codes are developed by these organizations:

1. Building Officials and Code Administrators International (BOCA). BOCA develops the *Basic/National Building Code*;
2. Southern Building Code Congress International (SBCCI). SBCCI develops the *Standard Building Code*; and
3. International Conference of Building Officials (ICBO). ICBO develops the *Uniform Building Code*.

Code requirements for fire alarm signaling systems generally define:

1. Occupancy descriptions and classifications.
2. Locations for smoke detectors.
3. Function of the fire alarm and emergency communication system (if an emergency communication system is determined necessary for the occupancy protected).
4. Operation of the voice/alarm function of a system.
5. Provision for a fire department communication system.
6. Components of the fire command station.
7. Emergency power requirements and service.
8. Manual fire alarm station location and use.
9. Exit door unlocking components.

TABLE 17-2. NFPA *101, Life Safety Code*, Requirements by Occupancy

	Hotels/Dormitories		Health Care	
	New	*Existing*	*New*	*Existing*
1. Detection, alarm and communications systems in accordance with Section 7-6	16-3.4.1	17-3.4.1	12-3.4.1.1 12-6.3.4.1	13-3.4.1 13-6.3.4.1
2. System smoke detectors in corridors, connected to the fire alarm system, installed per NFPA 72	16-3.4.4.1**			13-3.4.5.1¶
3. AC-powered single-station smoke detectors installed per NFPA 72				
a. In all dwelling units				
b. On each floor level (in all habitable rooms or in corridors)				
c. In each individual guest or sleeping room	16-3.4.4.2	17-3.4.4		
4. Manual pull stations, connected to the fire alarm system				
a. Within 200 ft of any point on any given floor	16-3.4.2	17-3.4.2‡	12-3.4.2§ 12-6.3.4.2	13-3.4.2§ 13-6.3.4.2
b. At a continuously supervised control point	16-3.4.2	17-3.4.2		
c. Near each required exit in the natural path of escape	16-3.4.2	17-3.4.2‡	12-3.4.2§ 12-6.3.4.2	13-3.4.2§ 13-6.3.4.2
5. Sprinkler supervisory equipment required to be connected to the fire alarm system	16-3.4.2	17-3.4.2	12-3.4.2 12-6.3.4.2	13-3.4.2 13-6.3.4.2
6. Occupant notification activated by the fire alarm system in accordance with Section 7-6.3	16-3.4.3.1	17-3.4.3.1	12-3.4.3.1 12-6.3.4.3	13-3.4.3.1 13-6.3.4.3
7. Voice communication system if high rise, in accordance with Section 30-8.3	16-4.1			
8. Central annunciator panel connected to the fire alarm system	16-3.4.3.2		12-3.4.3.3	
9. Fire alarm system connection to the fire department	16-3.4.3.3†	17-3.4.3.2†	12-3.4.3.2 12-6.3.4.4	13-3.4.3.2 13-6.3.4.4

* Applies to Class A mercantile only.
† Fire department notification by telephone permitted.
‡ Exempted if automatic initiation means is provided.
§ Exempted if manual pull stations provided at nurses' control stations.
¶ Required only in limited care facilities.
\# See exceptions.
** Exempted in sprinklered buildings.
For SI Units: 1 ft = 0.30 m.

TABLE 17-2. Life Safety Code, Requirements by Occupancy (continued)

Apartments		Mercantile		Business	
New	Existing	New	Existing	New	Existing
18-3.4.1#	19-3.4.1#	24-3.4.1* 24-4.4.3.1 24-4.5.3.1	25-3.4.1* 25-4.4.3.1 25-4.5.3.1	26-3.4.1#	27-3.4.1#
	Option 2 only				
18-3.4.4.1	19-3.4.4.1				
18-3.4.4.2					
18-3.4.2.1	19-3.4.2.1	24-3.4.2‡*	25-3.4.2‡*	26-3.4.2‡#	27-3.4.2‡#
18-3.4.2.1	19-3.4.2.1	24-3.4.2‡*	25-3.4.2‡*	26-3.4.2‡#	27-3.4.2‡#
18-3.4.2.2	19-3.4.2.4				
18-3.4.3.1	19-3.4.3.1	24-3.4.3.2 24-4.4.3.4 24-4.5.3.4	25-3.4.3.2 25-4.4.3.4 25-4.5.3.4	26-3.4.3.2	27-3.4.3.2
18-4.2				26-4.2	
18-3.4.3.2#	19-3.4.3.2#				
		24-3.4.3.3 24-4.4.3.5 24-4.5.3.4	25-4.4.3.5 25-4.5.3.5		

* Applies to Class A mercantile only.
† Fire department notification by telephone permitted.
‡ Exempted if automatic initiation means is provided.
§ Exempted if manual pull stations provided at nurses' control stations.
¶ Required only in limited care facilities.
# See exceptions.
** Exempted in sprinklered buildings.
For SI Units: 1 ft = 0.30 m.

EXAMPLE OF CODE REQUIREMENTS

The following is a composite of life safety provisions in model building codes and NFPA *101, Life Safety Code*. (NOTE: Occupancies in a community that are administered by the federal government are usually exempt from local code requirements.)

General

These recommendations cover basic functions of complete automatic fire alarm and emergency communication systems. These systems are primarily intended to provide early indication of emergency conditions.

Applicability

These recommendations apply to all occupancies of the following use groups, as typically defined by the building code in effect in the locality.

Use Group B, Businesses: "Businesses" are defined herein to include all buildings and structures, or parts thereof, that are used for the transaction of business, for the rendering of professional services, or for other services that involve stocks of goods, wares, or merchandise in limited quantity for incidental office use or sample purposes. "Businesses" is also meant to include offices, banks, civic administration facilities, outpatient clinics, professional buildings, testing and research laboratories, radio stations, telephone exchanges, and similar establishments.

Use Group R1, Residential and Hotel Occupancies: "Residential and hotel occupancies" are defined herein to include all hotel, motel, and dormitory buildings that are arranged for shelter and sleeping accommodations of more than 20 persons.

Use Group R2, Residential Multi-Family Occupancies: "Residential multi-family occupancies" are defined herein to include all multi-family dwellings having more than two dwelling units. This occupancy classification also includes dormitories, boarding, and lodging houses that are arranged for shelter and sleeping accommodations of more than five but not more than 20 persons.

When required by the authority having jurisdiction, the specified occupancies contain a listed fire alarm system for life safety. This system should be installed, tested, maintained, and used in accordance with applicable requirements. These requirements can be found in NFPA *101, Life Safety Code*.

Smoke Detection Systems

An approved smoke detection system is to be installed in the previously defined occupancies in the following locations:

1. Boiler and furnace rooms.
2. Return air ducts and plenums of heating, ventilating, and air conditioning (HVAC) systems serving floors other than the floor on which the HVAC equipment is located. Detectors should be located at each opening into the vertical return air ducts or shafts.
3. Corridor areas.
4. Elevator lobby areas.
5. Elevator penthouse areas.
6. Other areas that may be deemed necessary.

The detection system must be designed to activate the voice/alarm system on a selective basis. This selective activation must be dependent upon the compartmentation design of the system. The detection system, upon activation, must also place into operation all equipment necessary to prevent the spread of smoke.

In use groups R1 and R2 occupancies as defined by NFPA *101, Life Safety Code,* an approved single-station smoke detector must be installed in each room and residential dwelling.

Fire Alarm and Emergency Communication Systems

An approved fire alarm and emergency communication system must be installed for all occupancies in accordance with applicable requirements of NFPA *101, Life Safety Code.*

Voice/Alarm Systems

The operation of any smoke detector, sprinkler waterflow device, or manual fire alarm station in the occupancy must automatically activate a voice/alarm system. Activation of such system must automatically sound an alert signal to desired areas in the occupancy protected.

The voice/alarm system must provide a predetermined message on a selective basis to the area where the alarm originated. The message must also provide information and give directions to the occupants. The alarm must also be so designed as to be clearly heard by all hearing-able occupants within all designated areas.

Voice/alarm system controls must be located in the fire command station and operated from that station. Voice/alarm system controls should be located so that a selective or general voice alarm can be initiated by the station operator. Communication must be established on either a selective or general basis to the following terminal areas: elevators, elevator lobbies, corridors, exit stairways, rooms and tenant spaces exceeding 1,000 sq ft (92.936 m^2), dwelling units in apartment houses, and hotel guest rooms or suites.

Installation wiring for a voice/alarm system must be electrically supervised on a continuous basis. This supervision must detect opens, shorts, and ground conditions that could impair the function of the system.

Fire Department Communication System

A two-way fire department communication system must be provided for fire department use. This system must operate between the fire command station and every elevator, elevator lobby, and entry-enclosed exit stairway.

Fire Command Station

A fire command station for fire department operations must be provided in a location approved by the fire department. The fire command station must have the following components:

1. A voice alarm system panel.
2. A fire department communication panel.
3. Fire detection and alarm system annunciator panels.
4. Status indicators for elevators and annunciator. Such indicators must specify which elevators are operational.
5. Status indicators and controls for air-handling systems.
6. Controls for simultaneous unlocking of all stairway doors.
7. Sprinkler valve and waterflow detector display panels.
8. Status indicators for emergency power, light, and emergency system control.
9. A telephone for fire department use, which must have controlled access to the public telephone system.

Emergency Power

Emergency power provided must be capable of operating the emergency communication system during a fire or other emergency conditions. The emergency power must be derived by connection to an emergency power circuit of an emergency system.

NOTE: NFPA 70, *National Electrical Code*, describes emergency power circuit requirements and is used as the local electrical code in many localities.

The emergency power requirements must provide service to the following emergency systems: voice/alarm system, fire department communication system, fire department elevator, mechanical air-handling system, fire detection and alarm system, fire protection equipment and devices, exitway and other emergency lighting, and exitway door unlocking system.

Manual Fire Alarm Stations

Manual fire alarm stations must be provided and used only for fire protective signaling purposes. These stations must be provided in the natural path of escape near the required exit from each area of the occupancy protected. Additional manual fire alarm stations must be located within 200 ft (61 m) horizontal from any point in the building.

Doors

All exit stairway doors that are to be locked from the stairway side must have an approved lockset. This lockset must be able to be unlocked from the fire command station by the operator, and unlocked automatically upon actuation of the fire alarm system. The locks must be connected in a "fail-safe"-type manner so that, in event of a power failure, they will open automatically.

NFPA CODES AND STANDARDS

NFPA 72, *National Fire Alarm Code*, is the correlative standard to this text. However, other NFPA codes and standards with requirements applicable to a fire alarm signaling system exist, as listed in Table 17-3.

TABLE 17-3. NFPA Requirements for Fire Alarm Signaling Systems

NFPA Code or Standard	NFPA Requirements
NFPA 70	Wiring requirements for fire alarm signaling systems.
NFPA 101	System or protective functions required in various occupancies.
NFPA 110	Installation and performance of emergency generator equipment.
NFPA 1221	Municipal fire alarm and communication systems.

NOTE: Codes and standards specify minimum acceptable, real-world protection in an occupancy. A fire alarm signaling system designer is encouraged to exceed the minimum safety requirements in high-hazard occupancies, facilities with a heavy occupancy of transient or physically challenged occupants, or other facilities (and occupants therein) where extra protection is advised.

Further, the following NFPA standards contain requirements for automatic detection, alarm, and release of extinguishing agents that may also be part of a fire alarm signaling system.

NFPA 11, *Standard for Low-Expansion Foam and Combined Agent Systems.*

NFPA 11A, *Standard for Medium- and High-Expansion Foam Systems.*

NFPA 12, *Standard on Carbon Dioxide Extinguishing Systems.*

NFPA 12A, *Standard on Halon 1301 Fire Extinguishing Systems.*

NFPA 12B, *Standard on Halon 1211 Fire Extinguishing Systems.*

NFPA 13, *Standard for the Installation of Sprinkler Systems.*

NFPA 15, *Standard for Water Spray Fixed Systems for Fire Protection.*

NFPA 17, *Standard for Dry Chemical Extinguishing Systems.*

NFPA 17A, *Standard for Wet Chemical Extinguishing Systems.*

NFPA 75, *Standard for the Protection of Electronic Computer/Data Processing Equipment.*

NFPA 80, *Standard for Fire Doors and Fire Windows.*

NFPA 90A, *Standard for the Installation of Air Conditioning and Ventilating Systems.*

NFPA 170, *Standard for Firesafety Symbols.*

NFPA 231C, *Standard for Rack Storage of Materials.*

BIBLIOGRAPHY

NFPA Codes, Standards, Recommended Practices, and Manuals. (See the latest *NFPA Catalog* for availability of current editions of the following documents.)

NFPA 70, *National Electrical Code.*

NFPA 72, *National Fire Alarm Code.*

NFPA 101, *Life Safety Code.*

NFPA 1221, *Standard for the Installation, Maintenance, and Use of Public Fire Service Communication Systems.*

Additional Reading

See Tables 17-1 and 17-3, and additional NFPA codes and standards cited herein.

18

Household Fire Warning Systems

INTRODUCTION

Good fire alarm signaling system design is as necessary in residential occupancies — typically, one- and two-family homes — as it is in commercial structures. In 1992, for example, 74 percent of structural fires occurred in residential occupancies (72 percent in homes). Also in 1992, residential fires were responsible for 80 percent of all fire deaths, in structural or non-structural fires (78 percent in homes), and 47 percent of total dollar losses from fire (46 percent in homes). An effective household warning system integrates three elements: (1) minimizing fire hazards, (2) smoke detection equipment, and (3) a family escape plan. This chapter covers types of warning systems and detectors, and requirements for equipment, installation, maintenance, and testing. Additional detector information can be found in Chapter 4, "Signal Transmission"; Chapter 8, "Power Supplies"; and Chapter 14, "Testing."

HOUSEHOLD FIRE WARNING SYSTEM GUIDELINES

A household fire warning system is a system of devices that produces an audible alarm signal in the household to notify occupants of the presence of fire so they can evacuate the home.

Advance warning of a fire condition is critical to escape; actual fire tests in residential occupancies have shown that measurable amounts of smoke precede heat in almost all cases. Residential smoke detectors have lifesaving capabilities — but only if they are properly installed, regularly tested, and adequately maintained.

A typical household fire warning system includes individual single-station smoke detectors or a system of interconnected detectors. (Some ac systems are also connected to a local fire department.)

Residential smoke detectors were relatively expensive when they began to appear in the marketplace in the late 1960s. In 1970, however, the introduction of a battery-operated smoke detector, along with several new line-powered smoke detectors, initiated a period of public acceptance of single-station residential smoke detectors. Although heat detectors for residential use have been available since about 1921, field tests have shown they are less effective than smoke detectors in detecting fires in the home.

Current performance requirements and installation practices for residential fire detectors were developed from several test programs. The programs demonstrated that smoke detectors could provide a high level of life safety due to their fast response to a fire; and, secondly, detectable quantities of smoke usually preceded detectable levels of heat.

A study commonly known as the "Indiana Dunes Report," showed that at least one smoke detector on each story of a residence should be used as a minimum requirement. (This is the "every level" smoke detection system. A "story" is defined as the portion of a building between the upper surface of one floor and the upper surface of the floor or roof next above.) NFPA 72, *National Fire Alarm Code*, bases some of its fire detection criteria on this report.

TYPES OF SYSTEMS

A residential fire detection system can range in size from a single-station smoke detector in an apartment, mobile home, or recreational vehicle, to a group of multiple-station smoke detectors, or to a centrally wired system containing numerous detectors and separate alarm signaling devices. The fire detection system can include a burglar alarm system and an emergency medical alert system, and can be connected to a central receiving headquarters by leased telephone lines.

A direct-wired-type detector, when it is provided with one or more additional wires for interconnection with other similar detectors, is referred to as a multiple-station detector. (See Figure 18-1.) When interconnected according to instructions, an alarm from any one of the interconnected detectors sounds the alarm in all the detectors. This is important because the detector near the bedrooms should sound when a detector located in another area senses smoke. The number of detectors that can be interconnected varies from one model to another. Some interconnection methods also allow for connection of heat detectors or manual fire alarm boxes to the interconnecting wire. These multiple-station smoke detectors can be used to form a household system without installation of a separate control panel, but are limited to connections only within the individual household.

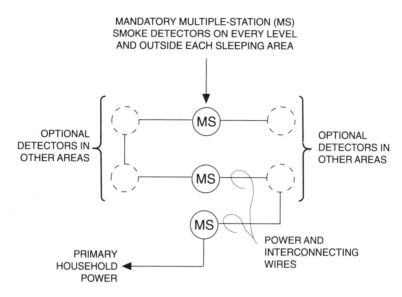

Figure 18-1. Typical household multiple-station smoke detector system.

Some smoke detectors are equipped with a transmitter that sends a signal to a receiver unit (usually located in a bedroom). If the detector is an ac plug-in model, a line carrier transmitter impresses a high-frequency signal on the house wiring. The receiver unit detects the signal as long as both are on the same power company transformer. If the detector is battery powered, it may contain a radio transmitter (similar to those used on garage door openers), with a range up to several hundred feet. A more detailed description of wireless transmission systems for residential and nonresidential use is found in Chapter 4, "Signal Transmission."

In addition to single- and multiple-station detectors, there are household fire warning systems similar in makeup and operation to a local fire alarm system. Such a system typically consists of a household system control unit that derives main power from the ac house wiring and usually contains a rechargeable standby battery capable of operating the system for at least 24 hours. (See Figure 18-2.) In addition, this system uses either smoke detectors of the system-connected type or single-station smoke detectors that contain alarm contacts. The fire warning system could also have heat detectors and manual fire alarm boxes, and separate alarm indicating appliances such as bells, horns, or electronic sirens. The control unit for a household warning system can also be connected to an automatic dialer, central station, or television cable to transmit the alarm to a point beyond the household. If a residential system is connected to some other alarm point, however, local code requirements for this type of alarm system must also be met.

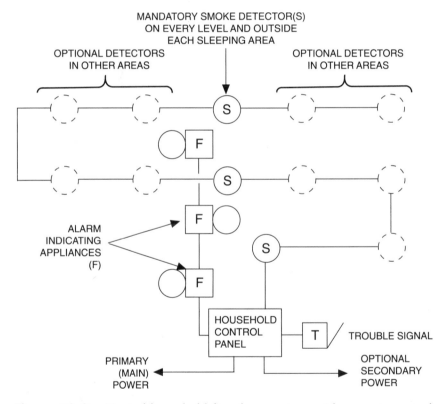

Figure 18-2. Typical household fire alarm system, with separate control panel.

Both wired and wireless household systems often include a burglar alarm in addition to fire alarm functions. The fire alarm signal must take precedence over the burglar alarm signal in combined systems with distinctive audible alarms so the homeowner can immediately discern the difference between a fire and a burglary. A medical alert feature is sometimes included on a system and provides for transmission of alarms to a point beyond the household via leased lines or television cables. (See Figure 18-3.)

* NOTE – MANY SYSTEM CONNECTED SMOKE DETECTORS
ARE AVAILABLE WITH BUILT-IN AUDIBLE SIGNAL.

Figure 18-3. Household fire alarm system.

The single- and multiple-station smoke detectors used in household fire warning systems operate on the same principle as the system-connected smoke detectors that are described in Chapter 3, "Signal Initiation." The main difference between them is that the single-station smoke detector is self-contained with a built-in alarm indicating appliance, while a multiple-station detector is basically a single-station detector with a method of inter-connecting all of the smoke detectors together so that, when one detector senses smoke, all of the alarm indicating appliances still sound. Typical single-station detectors are shown in Figure 18-4.

SINGLE STATION SMOKE DETECTORS

BATTERY OPERATED

IONIZATION

DETECTOR WITH
ESCAPE LIGHT

IONIZATION

PHOTOELECTRIC

AC POWERED

WITH HEAT DETECTOR

PHOTOELECTRIC

DETECTOR WITH STROBE LIGHT
FOR HEARING IMPAIRED

PHOTOELECTRIC

NOTE – MANY SINGLE STATION SMOKE DETECTORS ARE AVAILABLE WITH
CONNECTIONS FOR MULTIPLE STATION OPERATION.

Figure 18-4. Typical battery and ac-powered single-station smoke detectors.

A system-connected smoke detector does not usually have a built-in alarm appliance but (1) contains a relay that operates on alarm, (2) electronically impresses a signal on the alarm initiating circuit, or (3) sends the alarm signal to the control panel by wireless transmission. See Chapter 4, "Signal Transmission."

HOUSEHOLD WARNING SYSTEMS FOR ONE- AND TWO-FAMILY DWELLINGS

As defined by NFPA *101, Life Safety Code,* one- and two-family dwellings include buildings containing not more than two dwelling units in which each living unit is occupied by members of a single family with no more than three outsiders, if any, accommodated in rented rooms.

Detection

Approved single-station or multiple-station smoke detectors continuously powered by the house electrical service should be installed in accordance with NFPA 72, *National Fire Alarm Code.* Many codes allow battery-powered detectors to be used in existing construction, but require ac-powered detectors for new construction.

Single-station smoke detectors should be interconnected only within an individual living unit. Remote annunciation from single-station detectors can be permitted.

A control panel and associated equipment, multiple- or single-station alarm device(s), or any combination thereof can be used to form a household fire warning system in one- and two-family dwellings.

Ideally, smoke detectors should be installed outside of each separate sleeping area in the immediate vicinity of the bedrooms and on each additional story of the family living unit, including basements and excluding crawl spaces and unfinished attics. (See Figures 18-5 through 18-8.)

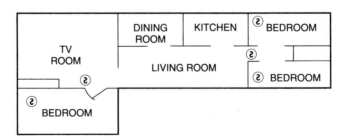

Figure 18-5. In family living units with more than one sleeping area, a smoke detector (indicated by cross) should be provided to protect each separate sleeping area. In new construction additional smoke detectors are required within each bedroom.

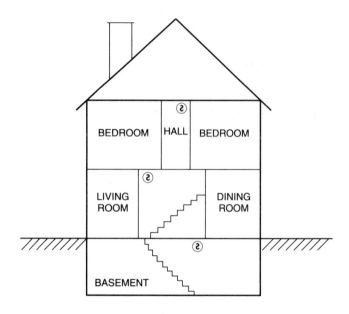

Figure 18-6. As a minimum a smoke detector should be located on each story.

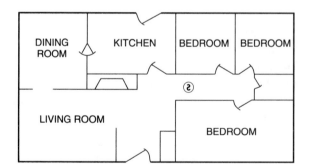

Figure 18-7. A smoke detector should be located between the sleeping area and the rest of the family living unit.

Smoke Detector Location, Spacing, and Installation

The major threat from fire in a family living unit is at night when everyone is asleep. The principal threat to persons in sleeping areas comes from fires in the remainder of the unit; therefore, smoke detec-tor(s) are best located between the bedroom areas and the rest of the

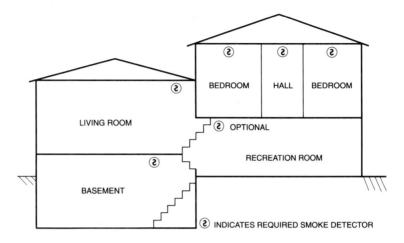

Figure 18-8. Required smoke detectors in a split-level arrangement. (The indicated smoke detector is optional if no door is provided between split-level living and recreation rooms.)

unit. In units with only one bedroom area on one floor, the smoke detector should be located as shown in Figure 18-7.

For family living units with one or more split levels (i.e., adjacent levels with less than one full story separation between levels), a smoke detector should suffice for an adjacent lower level, including basements. (See Figure 18-8.)

In new construction, the model building codes and NFPA 72, *National Fire Alarm Code*, now require smoke detectors within sleeping rooms. These provide bedroom occupants protection from a fire starting within that bedroom where the door to that bedroom is closed. A smoke detector outside the bedroom provides protection to others, but is ineffective for those in the room of origin. Further, since smoke detectors in new construction are required to be interconnected, so that when one alarms it sounds all of the detectors, the audibility of the entire system is enhanced along with the ability of the system to awaken even the soundest sleeper.

Certain guidelines should be followed for locations of smoke detectors in certain areas of the home. In rooms with ceiling slopes greater than 1 ft rise per 8 ft (0.3 m rise per 2.4 m) horizontally, smoke detectors should be located at the high side of the room. Smoke detectors should be located on or near the ceiling whenever practical. A smoke detector installed in a stairwell should be located so smoke rising in the stairwell cannot be prevented from reaching the detector by an inter vening door or obstruction. A smoke detector installed in the basement should be located in close proximity to the stairway leading to the floor above. A smoke detector installed to protect a sleeping area should be located outside of the bedrooms but in the immediate vicinity of the sleeping area. And, a smoke detector installed on a story without a separate sleeping area should be located in close proximity to the stairway leading to the floor above.

Smoke and heat detectors should be installed in those locations recommended by the manufacturer except when the space above the ceiling is open to the outside and little or no insulation is present over the ceiling. This results in ceiling temperatures that are cold in winter and hot in summer. Where the ceiling is significantly different in temperature from the air space below, smoke (and heat) has difficulty reaching the ceiling and to a detector that may be placed there. If a detector must be installed in these conditions, placement of the detector on a sidewall, with the top 4 to 12 in. (0.1 to 0.3 m) from the ceiling, is preferred. It should be recognized that the condition of inadequately insulated ceilings and walls can also exist in multi-family housing (apartments), single-family housing, and mobile homes. This situation can also exist (but to a lesser extent) with outside walls. While a detector should be installed optionally on the ceiling and then on a sidewall, if the sidewall is an exterior wall with little or no insulation, the detector should be installed on an interior wall.

Smoke detectors should be located no closer than 3 ft (0.9 m) from heating vents so that air issuing from the vent will not blow smoke away from the detector. Some smoke detectors are not suitable for location within kitchens because of false alarms from cooking vapors. Also, some smoke detectors are not recommended for garages (where automobile exhaust might cause alarms) or for attics or other unheated spaces where extremes of temperatures or humidity might affect detector operation. Before a smoke detector is installed in any of these locations, its specifications should be checked to ensure it is appropriate for the intended area.

In those family living units employing radiant heating in the ceiling, the wall location is preferred. Radiant heating in the ceiling can create a hot-air, boundary layer along the ceiling surface which can seriously restrict the movement of heat and smoke to a ceiling-mounted detector.

Maintenance and Testing

Household fire warning equipment should be maintained in accordance with the recommendations of the equipment manufacturer. In general, this means little more than keeping the equipment clean and free of dust, and replacing the batteries when needed. A maintenance agreement should be considered if the householder is unable to perform the required maintenance. NFPA 72, *National Fire Alarm Code*, requires that household systems (with control panels) be serviced by a trained professional at least every three years. Detectors should never be disconnected due to nuisance alarms (e.g., from cooking). The detector should be relocated or replaced with a detector less sensitive to cooking. A detector's sensitivity adjustment is normally set at the factory and should not be reset by an individual once the detector is in use.

Tests or inspections as recommended by the manufacturer should be performed by the homeowner at least once a month (for other than battery-powered detectors) and a minimum of once a week for battery-powered detectors. It is also a good practice to establish a specific schedule for these tests. The importance of regular tests of smoke detectors and other electronic fire alarm equipment cannot be overstressed. Further information on testing fire alarm equipment is found in Chapter 14, "Testing."

Proper care and maintenance of the detectors is equally as important as the installation. Each residential smoke detector is sold with an owner's booklet that describes the necessary maintenance procedures. Fire reports involving death or serious injury where smoke detectors were installed have shown in almost all cases that the detectors were inoperative at the time of the fire because the homeowner had (1) failed to replace a worn out battery, (2) failed to install the detector properly, (3) intentionally disconnected the power due to false alarms, or (4) failed to properly evacuate the dwelling. Because of this problem with battery-powered smoke detectors, NFPA 72, *National Fire Alarm Code*, mandates use of an ac primary power source for detectors in all new construction. This is a good recommendation even for existing smoke detectors in older construction.

SOUNDING APPLIANCES

Each detection device should cause the operation of an alarm. Alarm sound should be clearly audible in all bedrooms over background noise levels with all intervening doors closed. Audibility tests should be conducted under worst-case conditions (at night, with window air conditioners and room humidifiers operating). A single-station smoke detector sounds an audible alarm at the device itself, and sound level output should be a minimum of 85 dBA at a distance of 10 ft (3 m).

For instance, there may be a noisy window air conditioner or room humidifier, which may generate an ambient noise level of 55 dBA or higher. The detection devices' alarms must be able to penetrate through the closed doors and be heard over the bedroom's noise levels with sufficient intensity to awaken sleeping occupants therein. Test data indicate that detection devices having sound pressure ratings of 85 dBA at 10 ft (3 m) and installed outside the bedrooms can produce about 15 dBA over ambient noise levels of 55 dBA (approximately the minimum noise level of an air conditioner or other device) in the bedrooms.

Test studies have shown that detectors located remote from the bedroom area may not be loud enough to awaken the average person. In such cases, it is recommended that detectors be interconnected so that operation of the remote detector will cause an alarm of sufficient intensity to penetrate the bedrooms. This interconnection can be accomplished by installation of a fire detection system, wiring together multiple-station alarm devices, use of line carrier or radio frequency transmitters/receivers, etc.

Because an alarm may not be heard from a remote smoke detector, NFPA 72, *National Fire Alarm Code*, and many building codes now require that, where more than one detector is required to be installed within an individual dwelling unit, the detectors should be wired in such a manner that the actuation of one alarm will actuate all the alarms. This ensures that smoke anywhere in the dwelling unit will activate the detector(s) near the bedroom(s).

A wired or wireless household system normally has an alarm indicating appliance circuit to connect various alarm signals, such as fire alarm bells, horns, or electronic sirens. These appliances can produce higher sound levels and can be located independent of the detectors and can be clearly audible throughout the house (or even outside, where neighbors might hear the alarm).

VISIBLE APPLIANCES

NFPA 72, *National Fire Alarm Code*, requires that homes occupied by one or more hearing-impaired persons be equipped with flashing lights or approved tactile (vibrating) devices. The light intensities are specified on the basis of research conducted by Underwriters Laboratories Inc. (UL) on a large number of hearing-impaired persons. They found that a light intensity of 15 candela was sufficient to attract the attention of an awake person even if that person was not looking at the light. Further, NFPA 72 requires that the "every level" detectors be equipped with

15-candela lights. Because of their low power consumption, these lights should be operated from a source of standby power when provided.

The UL study found that significantly more intense light is required to awaken a sleeping person — 110 candela minimum for about a 95 percent waking efficiency. Thus, rooms in which hearing-impaired persons sleep are required to be provided with a 110-candela light mounted on the wall, 24 in. (0.6 m) below the ceiling, where the light will not be attenuated by any smoke collected at the ceiling. If mounted on the ceiling, the light intensity must be 177 candela to provide the 110 candela at the pillow surface, based on the maximum expected smoke density at the ceiling necessary to activate the smoke detector. These high-intensity lights cannot be powered practically from a battery supply, so such is not required.

One important note is that NFPA 72, *National Fire Alarm Code*, places the responsibility for the provision of these lights on the homeowner or tenant, because it is often not possible for a landlord or builder to determine that a person is hearing impaired.

A HOME FIRESAFETY "SYSTEM"

Minimizing Fire Hazards

Since it is always easier to prevent a fire rather than control one, minimizing fire hazards in and around the home is a key element to home firesafety and an asset to a home fire warning system. Minimizing hazards includes such things as removing all nonessential flammable liquids and other highly combustible items from the home, properly storing essential combustibles away from ignition sources, and not overloading electrical circuits. If a fire develops rapidly enough, however, residential smoke detectors may not provide sufficient time for the occupants of a dwelling to escape before exit routes become unreachable.

Escape Planning

Another key element in residential fire protection is developing and practicing a family escape plan. Residential smoke detectors can only warn occupants of a fire in the home. There are three important elements of a home escape plan that should be followed by all occupants of a home:

1. All occupants should leave the home immediately and call the fire department from a neighbor's house. Many deaths have been reported in fires where people have unwisely taken extra time to get dressed, gather valuables, or look for pets.

2. Everyone should have an alternate way out of each room in case the primary exit is blocked by fire.

3. A prearranged outside meeting place should be determined so everyone will know that the entire family has escaped the fire. Deaths have resulted because people have reentered a burning home to look for persons who had already escaped.

BIBLIOGRAPHY

NFPA Codes, Standards, Recommended Practices, and Manuals. (See the latest *NFPA Catalog* for availability of current editions of the following documents.)

NFPA 72, *National Fire Alarm Code.*

NFPA *101, Life Safety Code.*

Additional Reading

Bukowski, R. W., Waterman, T. E., and Christian, W. J., "Detector Sensitivity and Siting Requirements for Buildings," 1975, National Bureau of Standards, Washington, DC.

"Protecting Your Family from Fire," FA 130 (in Spanish, FA 129), US Fire Administration, Gaithersburg, MD.

Appendix

FIRE ALARM AND EMERGENCY COMMUNICATION SYMBOLS

REFERENT (SYNONYM)	SYMBOL	COMMENTS

SIGNAL INITIATING DEVICES

REFERENT (SYNONYM)	SYMBOL	COMMENTS
MANUAL STATIONS (CALL POINT)	▢	GENERAL
MANUAL ALARM BOX (PULL STATION AND PULL BOX)	▣	
TELEPHONE STATION (TELEPHONE CALL POINT)		
AUTOMATIC DETECTION AND SUPERVISORY DEVICES	◯	GENERAL
HEAT DETECTOR* (THERMAL DETECTOR)		INCLUDES FIXED TEMPERATURE, RATE COMPENSATION, AND RATE-OF-RISE DETECTORS
SMOKE DETECTOR		INCLUDES PHOTOELECTRIC AND IONIZATION-TYPE DETECTORS
SMOKE DETECTOR IN DUCT		
GAS DETECTOR	●	
FLAME DETECTOR* (FLICKER DETECTOR)		INCLUDES ULTRAVIOLET, INFRARED AND VISIBLE RADIATION-TYPE DETECTORS
FLOW DETECTOR/ SWITCH		
PRESSURE DETECTOR/ SWITCH*		ALTERNATE TERM: PRESSURE SWITCH — AIR, WATER, ETC.
LEVEL DETECTOR/ SWITCH*		
TAMPER DETECTOR/ SWITCH		ALTERNATE TERM: TAMPER SWITCH
VALVE WITH TAMPER DETECTOR/SWITCH		

* SYMBOL ORIENTATION MUST NOT BE CHANGED.

FIRE ALARM AND EMERGENCY COMMUNICATION SYMBOLS (CONT.)

REFERENT (SYNONYM)	SYMBOL	COMMENTS
AUDIBLE-TYPE ALERTING DEVICES (SOUNDER)		
SPEAKER/HORN (ELECTRIC HORN)		
BELL (GONG)		
WATER MOTOR ALARM (WATER MOTOR GONG)		SHIELD OPTIONAL
HORN WITH LIGHT (HORN WITH STROBE)		
VISUAL TYPE		
LIGHT (LAMP, SIGNAL LIGHT, INDICATOR LAMP, STROBE)		
ILLUMINATED EXIT SIGN		
ILLUMINATED EXIT SIGN WITH DIRECTION ARROW		
EMERGENCY ILLUMINATION SYMBOLS		
EMERGENCY LIGHT, BATTERY POWERED, ONE LAMP		
EMERGENCY LIGHT, BATTERY POWERED, TWO LAMPS		
EMERGENCY LIGHT, BATTERY POWERED, THREE LAMPS		
CONTROL AND SUPERVISORY DEVICES		
CONTROL PANEL		
DOOR HOLDER		

WATER SUPPLY AND DISTRIBUTION SYMBOLS

REFERENT (SYNONYM)	SYMBOL	COMMENTS
MAINS, PIPE		
PUBLIC WATER MAIN	————————	INDICATE PIPE SIZE
PRIVATE WATER MAIN	════════	INDICATE PIPE SIZE
WATER MAIN UNDER BUILDING	– – – – – –	INDICATE PIPE SIZE
SUCTION MAIN	▬ ▬ ▬	INDICATE PIPE SIZE
VALVES		
POST INDICATOR AND VALVE	⊢--▷◁--⊣	INDICATE VALVE SIZE
KEY-OPERATED VALVE	⊢--▷◁--⊣	INDICATE VALVE SIZE
OS & Y VALVE (OUTSIDE SCREW AND YOKE, RISING STEM)	⊢--▷◁--⊣	
INDICATING BUTTERFLY VALVE	⊢-----▷◁-----⊣	INDICATE VALVE SIZE
NONINDICATING VALVE (NONRISING-STEM VALVE)	⊢--▷◁--⊣	
VALVE IN PIT	⊢-◁▷-⊣	INDICATE VALVE SIZE
CHECK VALVE	⊢-◁-⊣ →	INDICATE VALVE SIZE

Glossary

Alarm Service. The service required following the receipt of an alarm signal.

Alarm Signal. A signal indicating an emergency requiring immediate action, as an alarm for fire from a manual box, a waterflow alarm, an alarm from an automatic fire alarm system, or other emergency signal.

Alarm System. A combination of compatible initiating devices, control panels, and indicating appliances designed and installed to produce an alarm signal in the event of fire.

Annunciator. A unit containing two or more identified targets or indicator lamps in which each target or lamp indicates the circuit, condition, or location to be annunicated.

Authority Having Jurisdiction. The organization, office, or individual responsible for "approving" equipment, an installation, or a procedure.

NOTE: The phrase "authority having jurisdiction" is used in NFPA documents in a broad manner since jurisdictions and "approval" agencies vary as do their responsibilities. Where public safety is primary, the "authority having jurisdiction" may be a federal, state, local, or other regional department or individual such as a fire chief, fire marshal, chief of a fire prevention bureau, labor department, health department, building official, electrical inspector, or others having statutory authority. For insurance purposes, an insurance inspection department, rating bureau, or other insurance company representative may be the "authority having jurisdiction." In many circumstances the property owner or a designated agent assumes the role of the "authority having jurisdiction"; at government installations, the commanding officer or departmental official may be the "authority having jurisdiction."

Auxiliarized Local System. A local system that is connected to the municipal alarm facilities.

Auxiliarized Proprietary System. A proprietary system that is connected to the municipal alarm facilities.

Auxiliary Protective Signaling System. A connection to the municipal fire alarm system to transmit an alarm of fire to the municipal communication center. Fire alarms from an auxiliary alarm system are received at the municipal communication center on the same equipment and by the same alerting methods as alarms transmitted from municipal fire alarm boxes located on streets.

Auxiliary Trip Relay. A relay used to operate a municipal master box from an auxiliarized control panel.

Bell, Single Stroke. A bell whose gong is struck only once each time operating energy is applied.

Bell, Vibrating. A bell that rings continuously as long as operating power is applied.

Box (or Station), Fire Alarm.

1. **Noncoded.** A manually operated device that, when operated, closes or opens one or more sets of contacts and generally locks the contacts in the operated position until the box is reset.

2. **Coded.** A manually operated device in which the act of pulling a lever causes the transmission of not fewer than three rounds of coded alarm signals. Similar to the noncoded type; except that, instead of a manually operated switch, a mechanism to rotate a code wheel is utilized. Rotation of the code wheel, in turn, causes an electrical circuit to be alternately opened and closed, or closed and opened, thus sounding a coded alarm that identifies the location of the box. The code wheel is cut for the individual code to be transmitted by the device and can operate by clockwork or by an electric motor. Clockwork transmitters can be prewound or can be wound by the pulling of the alarm lever. Usually the box is designed to repeat its code four times and automatically come to rest. Prewound transmitters must sound a trouble signal when they require rewinding. Solid-state electronic coding devices are also used in conjunction with the fire alarm control unit to produce coded sounding of the audible signaling appliances.

Breakglass Box (or Station). A breakglass box is one in which it is necessary to break a special element in order to operate the box.

Ceiling. The upper surface of a space, regardless of height. Areas with a suspended ceiling would have two ceilings, one visible from the floor and one above the suspended ceiling.

Ceiling Height. The height from the continuous floor of a room to the continuous ceiling of a room or space.

Central Station System. A system, or group of systems, in which the operations of circuits and devices are signaled automatically to, recorded in, maintained, and supervised from an approved central station having competent and experienced observers and operators who, upon receipt of a signal, take the required action. Such systems are controlled and operated by a person, firm, or corporation whose principal business is the furnishing and maintaining of supervised signaling service.

Channel. A path for signal transmission between two or more stations or channel terminations. A channel can consist of wire, radio waves, or equivalent means of signal transmission.

Chimes. A single-stroke or vibrating-type audible signal appliance that has a xylophone-type striking bar.

Circuit. The conductors or radio channel, and associated equipment, used to perform a definite function in connection with an alarm system.

Circuit Interface. A functional assembly that interfaces one or more of its initiating device circuits with a signaling line circuit in a manner that permits the central supervising station to indicate the status of each of its individual initiating device circuits.

Coded Signal. A signal pulsed in a prescribed code for each round of transmission. A minimum of three rounds and a minimum of three impulses is required for an alarm signal.

Combination Detector. A device that either: (1) responds to more than one of the fire phenomena, such as smoke, heat, flame, and fire gas or (2) employs more than one operating principle to sense one of these phenomena. Typical examples are: (1) a combination of a heat detector with a smoke detector or (2) a combination rate-of-rise and fixed temperature heat detector.

Combination System. A local protective signaling system for fire alarm, supervisory, or guard's tour supervisory service whose components may be used in whole or in part in common with a nonfire signaling system, such as a paging system, a burglar alarm system, a musical program system, or a process monitoring service system, without degradation of or hazard to the protective signaling system.

Communication Channel. A signaling channel (usually leased from a communication utility company) having two or more terminal locations and a suitable information handling capacity depending on the characteristics of the system used. One terminal location is at the central supervising station and the other terminal location or locations are sources from which are transmitted alarm signals, supervisory signals, trouble signals, and such other signals as the central supervising station is prepared to receive and interpret.

Control Unit. A device with the control circuits necessary to: (1) furnish power to a fire alarm system, (2) receive signals from alarm initiating devices and transmit them to audible alarm indicating appliances and accessory equipment, and (3) electrically supervise the system installation wiring and primary (main) power. The control unit can be contained in one or more cabinets in adjacent or remote locations.

Delinquency Signal. A signal indicating the need for action in connection with the supervision of guards or system attendants.

Emergency Voice/Alarm Communication System. A system that provides dedicated manual or automatic, or both, facilities for originating and distributing voice instructions, as well as alert and evacuation signals pertaining to a fire emergency to the occupants of a building.

End-of-Line Device. A device used to terminate a supervised circuit.

Fault. An open, ground, or short condition on any line(s) extending from a control unit, which could prevent normal operation.

Flame Detector. A device that detects the infrared, or ultraviolet, or visible radiation produced by a fire.

Frequency Division Multiplexing. A signaling method characterized by the simultaneous transmission of more than one signal in a communication channel. Signals from one or multiple terminal locations are distinguished from one another by virtue of each signal being assigned to a separate frequency or combination of frequencies.

Ground Fault. A condition in which the resistance between a conductor and ground reaches an unacceptably low level.

Ground Fault Detector. Detects the presence of a ground condition on system wiring.

Heat Detector. A device that detects abnormally high temperature or rate-of-temperature rise.

Horns. An audible signal appliance in which energy produces a sound by imparting motion to a flexible component that vibrates at some nominal frequency.

Indicating Appliance. Any audible or visible signal employed to indicate a fire, supervisory, or trouble condition. Examples of audible signal appliances are bells, horns, sirens, electronic horns, buzzers, and chimes. A visible indicator consists of a lamp, target, meter deflection, or equivalent.

Indicating Appliance Circuit. A circuit or path directly connected to an indicating appliance(s), such as bells, horns, chimes, etc.

Initiating Device (Appliance). A manually or automatically operated device, the normal intended operation of which results in a fire alarm or supervisory signal indication from the control unit. Examples of alarm signal initiating devices are thermostats, manual boxes, smoke detectors, and waterflow switches. Examples of supervisory signal initiating devices are water-level indicators, sprinkler-system valve-position switches, pressure supervisory transmitters, and water temperature switches.

Initiating Device Circuit. A circuit to which automatic or manual signal-initiating devices, such as fire alarm boxes, fire detectors, and waterflow alarm devices, are connected.

Labeled. Equipment or materials to which has been attached a label, symbol or other identifying mark of an organization acceptable to the "authority having jurisdiction" and concerned with product evaluation, that maintains periodic inspection of production of labeled equipment or materials and by whose labeling the manufacturer indicates compliance with appropriate standards or performance in a specified manner.

Leg Facility. That part of a signaling line circuit connecting each protected building to the trunk facility or directly to the central supervising station.

Listed. Equipment or materials included in a list published by an organization acceptable to the "authority having jurisdiction" and concerned with product evaluation, that maintains periodic inspection of production of listed equipment or materials and whose listing states either that the equipment or material meets appropriate standards or has been tested and found suitable for use in a specified manner.

NOTE: The means for identifying listed equipment may vary for each organization concerned with product evaluation, some of which do not recognize equipment as listed unless it is also labeled. The "authority having jurisdiction" should utilize the system employed by the listing organization to identify a listed product.

Local Alarm System. A local system sounding an alarm as the result of the manual operation of a fire alarm box or the operation of protection equipment or systems, such as water flowing in a sprinkler system, the discharge of carbon dioxide, the detection of smoke, or the detection of heat.

Local Energy Auxiliary Alarm System. An auxiliary alarm system that employs a locally complete arrangement of parts, initiating devices, relays, power supply, and associated components to automatically trip a municipal transmitter or master box over electric circuits that are electrically isolated from the municipal system circuits.

Local Supervisory System. A local system arranged to supervise the performance of guards' tours, or the operative condition of automatic sprinkler systems, or other systems for the protection of life and property against the fire hazard.

Local System. A local system is one that produces a signal at the premises protected.

Maintenance. Repair service, including periodic inspections and tests, required to keep the protective signaling system and its component parts in an operative condition at all times, together with replacement of the system and its components, when for any reason they become undependable or inoperative.

Master Box. A municipal fire alarm box that may also be operated by remote means.

Multiplexing. A signaling method characterized by the simultaneous or sequential transmission, or both, and reception of multiple signals in a communication channel, including means for positively identifying each signal.

Municipal Communication Center. The building or portion of a building used to house the central operating part of the fire alarm system; usually the place where the necessary testing, switching, receiving, retransmitting, and power supply devices are located.

Municipal Fire Alarm Box. A specially manufactured enclosure housing a manually operated transmitter used to send an alarm to the municipal communication center.

Municipal Transmitter. A specially manufactured enclosure housing a transmitter that can only be tripped remotely, used to send an alarm to the municipal communication center.

Noncoded Signal. Signal from any indicating appliance that is energized continuously.

Paging System. A system intended to page one or more persons by means such as voice over loudspeaker stations located throughout the premises, or by means of coded audible signals or visual signals similarly distributed, or by means of lamp annunciators located throughout the premises.

Parallel Telephone Auxiliary Alarm System. An auxiliary alarm system connected by a municipally controlled individual circuit to the protected property, to interconnect the actuating devices and the municipal fire alarm switchboard.

Parallel Telephone System. A telephone system in which an individual wired circuit is used for each box.

Permanent Visual Record (Recording). Immediately readable, not easily alterable print, slash, punch, etc., listing all occurrences of status change.

Proprietary Protective Signaling System. An installation of protective signaling systems that serves contiguous and noncontiguous properties under one ownership from a central supervising station located at the protected property, where trained, competent personnel are in constant attendance. This includes the central supervising station; power supplies; signal-initiating devices; initiating device circuits; signal notification appliances; equipment for the automatic, permanent visual recording of signals; and equipment for the operation of emergency building control services.

Protective Signaling Systems. Electrically operated circuits, instruments, and devices, together with the necessary electrical energy, designed to transmit alarm, supervisory, and trouble signals necessary for the protection of life and property.

Rectifier. An electrical device without moving parts that changes alternating current (ac) to direct current (dc).

Remote Station Protective Signaling System. An installation using supervised dedicated circuits, installed to transmit alarm, supervisory, and trouble signals from one or more protected premises to a remote location at which appropriate action is taken.

Repeater Facility. Equipment needed to relay signals between the protected premises and the central supervising station.

Runner Service. Employees, other than the required number of operators, on duty at all times at the central supervising station, at a runner station, or in a vehicle in constant radio contact with the central supervising station, available for prompt dispatching, when necessary, to the protected premises.

Shunt Auxiliary Alarm System. An auxiliary alarm system electrically connected to an integral part of the municipal alarm system extending the municipal circuit into the protected property to interconnect the actuating devices. When operated, these devices open the municipal circuit shunted around the trip coil of the municipal transmitter or master box, which is thereupon energized to start transmission without any assistance whatsoever from a local source of energy.

NOTE: The shunt system runs municipal power wires into protected premises. Thus, the municipality may lose control of its circuit. In addition, an open circuit in this shunt loop will cause an alarm condition. The use of a shunt-type system is a matter of individual municipal policy.

Signaling Line Circuit (Path). A circuit or path (channel or trunk and leg) over which multiple signals are transmitted and received.

Smoke Detector. A device that detects visible or invisible particles of combustion.

Spacing. A horizontally measured dimension relating to the allowable coverage of fire detectors.

Supervision. Refers to monitoring of the circuit, switch, or device in such a manner that a trouble signal is received when a fault that would prevent normal operation of the system occurs.

Supervisory Service. The service required to monitor performance of guard patrols and the operative condition of automatic sprinkler systems and of other systems for the protection of life and property.

Supervisory Signal. A signal indicating the need of action in connection with the supervision of guards' tours, sprinkler and other extinguishing systems or equipment, or with the maintenance features of other protective systems.

Supplementary. Refers to equipment or operation not required by signaling system standards and designated as such by the authority having jurisdiction.

Transmitter. A system component to which initiating devices or groups of initating devices are connected. The component transmits signals to the central supervising station indicating the status of the initiating devices and the initiating device circuits.

Trouble Signal. An audible signal indicating trouble of any nature, such as a circuit break or ground, occurring in the devices or wiring associated with a protective signaling system.

Trunk Facility. That part of a signaling line circuit connecting two or more leg facilities to the central supervising station or satellite station.

Two-Way Fire Department Communication System. An electrically supervised telephone system providing private voice communication capability between the command center or central control panel and designated remote locations.

Visible Signal. A visible signal is the response to the operation of an initiating device by one or more direct or indirect visible notification appliances. For a direct visible signal, the sole means of notification is by direct viewing of the light source. For an indirect visible signal, the sole means of notification is by illumination of the area surrounding the visible signaling appliance.

Waterflow Switch. An assembly approved for the service and so constructed and installed that any flow of water from a sprinkler system equal to or greater than that from a single automatic sprinkler of the smallest orifice size installed on the system will result in activation of this switch and subsequently indicate an alarm condition.

Zone. A designated area of a building. Commonly, zones within a building are annunciated to rapidly locate a fire.

Index